Aberrations of Optical Systems

The Adam Hilger Series on Optics and Optoelectronics

Series Editors: **E R Pike** FRS and **W T Welford** FRS

Other books in the series

Laser Damage in Optical Materials
R M Wood

Waves in Focal Regions
J J Stamnes

Laser Analytical Spectrochemistry
edited by V S Letokhov

Laser Picosecond Spectroscopy and Photochemistry of Biomolecules
edited by V S Letokhov

Prism and Lens Making
F Twyman

Infrared Optical Fibers
T Katsuyama and H Matsumura

Solar Cells and Optics for Photovoltaic Concentration
A Luque (to be published in 1989)

The Fabry–Perot Interferometer: History, Theory, Practice and Applications
J M Vaughan (to be published in 1989)

Other titles of related interest

Infrared Optical Materials and their Antireflection Coatings
J A Savage

Principles of Optical Disc Systems
G Bouwhuis, J Braat, A Huijser, J Pasman, G van Rosmalen and
K Schouhamer Immink

Thin-film Optical Filters
H A Macleod

Aberration and Optical Design Theory
G G Slyusarev

Lasers in Applied and Fundamental Research
S Stenholm

The Optical Constants of Bulk Materials and Films
L Ward

The Adam Hilger Series on Optics and Optoelectronics

Aberrations of Optical Systems

W T Welford FRS

The Blackett Laboratory,
Imperial College of Science and Technology,
University of London

Adam Hilger, Bristol and Philadelphia

British Library Cataloguing in Publication Data

Welford, W.J.
    Aberrations of optical systems.
    1. Optics, Geometrical   2. Aberration
    I. Title
    535'.32   QC381

    ISBN 0-85274-564-8

Library of Congress Cataloging-in-Publication Data are available

First published 1986
Reprinted with minor corrections 1989

Published under the Adam Hilger imprint by IOP Publishing Ltd
Techno House, Redcliffe Way, Bristol BS1 6NX, England
242 Cherry Street, Philadelphia, PA 19106, USA

Printed in Great Britain by J W Arrowsmith Ltd, Bristol

Contents

Preface and Acknowledgements

Aberration theory as used in optical design has changed considerably since 1974 when my book "Aberrations of the Symmetrical Optical System" (Academic Press) appeared. Among these changes are the use of non-axially symmetric systems and diffractive optical elements in quite complex designs such as head-up displays and the increasing use of scanning systems with laser illumination. The present book is based to a considerable extent on that of 1974 and I acknowledge the use of much material from that book; however, I have changed much and added material on the subjects mentioned above and others suggested by colleagues. It is a pleasure to acknowledge help and suggestions given by these friends, including Prudence Wormell, Richard Bingham, Charles Wynne, Michael Kidger and many others. I am grateful also to Jim Revill of IOP Publishing Ltd for his advice and help in the preparation of the book.

W T Welford

1 Optical Systems and Ideal Optical Images

The symmetrical optical system, i.e. a system with symmetry about an axis of revolution, is the type of system most frequently met as a design problem; this includes systems folded by means of plane mirrors or prisms, since it is trivial to unfold them for optical design purposes. However, non-symmetrical systems are not uncommon, e.g. some kinds of spectacle lens, spectrographic systems, anamorphic projection systems and systems containing holographic optical elements. In this book we shall be mainly concerned with symmetrical systems but some discussion of non-symmetrical systems will be given, chiefly in connection with raytracing.

1.1 Initial assumptions

The treatment will be based mainly on the geometrical optics model but there will be occasional references to physical optics in the form of scalar wave theory; this is needed for dealing with aberration tolerances. In geometrical optics the essential concept is the ray of light; in this chapter we assume this as an intuitive notion, deferring more precise definition to Chapter 2. It is then possible to formulate definitions of ideal image formation using only the concept of rays and the assumption that to one ray entering the system there corresponds one and only one ray emerging. We do not at this stage invoke the laws of reflection and refraction, and we make no assumptions about how the transformation from object to image space is accomplished: i.e. there might be non-spherical surfaces, media of continuously varying refractive index, etc., in the system. However, it is convenient to assume that the entering and emerging rays are straight line segments, or in physical terms that there are clearly defined regions in the object and image spaces in which the respective refractive indices are constant. Ideal image formation for a general system then means that a pencil of rays from a point in object space becomes a pencil also passing through a single point in image space and that this holds for some one- or two-dimensional object surface. This does not get us very far but if we assume a symmetrical system we can obtain many other properties of ideal image formation to which the performance of a real well-corrected system should approximate.

The first notions of ideal image formation through symmetrical systems are due to A. F. Möbius (1855). A few years later James Clerk Maxwell (1856, 1858) formalized the concept of an ideal system without invoking any physical image-forming mechanism. It is essentially Maxwell's concept which we describe in this chapter.

1.2   Ideal image formation in the symmetrical optical system

Take the z-axis of a right-handed Cartesian coordinate system as the axis of revolution of a symmetrical optical system, as in Fig. 1.1, and the y-axis in the plane of the diagram: the origin O is taken as any convenient point on the axis.

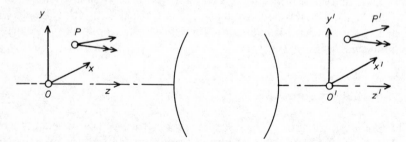

Fig. 1.1. Coordinates for the symmetrical optical system

If, as is customary, the light is supposed to travel from left to right then this coordinate system is in the object space and we take a similar system $O'x'y'z'$ in the image space, the respective axes being parallel to each other.

All points and rays in object space are referred to $Oxyz$ and those in image space to $O'x'y'z'$. The rays are shown as straight line segments but we introduce immediately the generalization that they are to be regarded as extending indefinitely in either direction; thus the object space extends right through the optical system and through image space and similarly image space is extended infinitely in both directions. This is an essential convention for dealing with the details of image formation in the intermediate spaces of a system, where the image from one optical element is the object for the next, since this image-*cum*-object is very frequently virtual.

Ideal image formation from the x–y plane to the x'–y' plane can then be defined as follows. All rays through any point P on the x–y plane must pass through one point P' on the x'–y' plane and the coordinates (x', y') are proportional to (x, y); the constant of proportionality, which is, of course, the magnification, depends on the nature of the optical system and on the axial positions of the two planes. This can be summarized by saying that any figure on the x–y plane is perfectly imaged as a geometrically similar figure on the x'–y' plane. It will be shown that if there are two such pairs of conjugate planes then any plane in object space is imaged ideally on another plane in image space.

There is considerable interest in examining this Maxwellian ideal image formation because, as will be seen in Chapter 3, the image formation in any real symmetrical system approximates to the ideal in a narrow region sufficiently close to the optical axis. We shall therefore study this ideal image formation in more detail.

Let Oxy and $O_1x_1y_1$ be two planes in object space and let $O'x'y'$ and $O'_1x'_1y'_1$ be the corresponding planes in image space. We suppose the image formation to be perfect between these pairs of planes; thus, in Fig. 1.2, if a

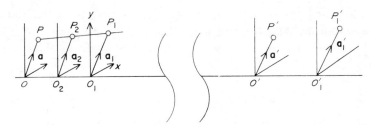

FIG. 1.2. Ideal image formation

and a_1 are two-dimensional vectors in the planes Oxy and $O_1x_1y_1$ from the origins to points P and P_1, the corresponding vectors in the image planes will be given by

$$\left.\begin{array}{l} a' = ma \\ a'_1 = m_1 a_1 \end{array}\right\} \tag{1.1}$$

where m and m_1 are appropriate magnification factors between these planes. These equations imply that *all* rays through P pass through P′ and similarly for P_1 and P_1', so that they express the assumption of perfect image formation between the two planes.

Now we consider points P_2 on a third plane $O_2 x_2 y_2$; we enquire whether all rays through P_2 also pass through one point P_2' in image space and, if so, whether this point always lies on one plane perpendicular to the axis and is related to P_2 by an equation similar to eqn (1.1). We have

$$a_2 = (z_2/z_1)(a_1 - a) + a \qquad (1.2)$$

where z_1 and z_2 are the coordinates of O_1 and O_2 with respect to O, this being true for all points P and P_1 which are on a ray through P_2. If a point P_2' with the above properties exists, then we must also have

$$a_2' = (z_2'/z_1')(a_1' - a') + a' \qquad (1.3)$$

and

$$a_2' = m_2 a_2. \qquad (1.4)$$

Equations (1.3) and (1.2) can be rearranged as

$$a_2' = (m_1 z_2'/z_1')a_1 + m(1 - z_2' / z_1')a \qquad (1.3')$$

$$a_2 = (z_2/z_1)a_1 + (1 - z_2/z_1)a \qquad (1.2')$$

and these will be consistent with eqn (1.4) if we can find z_2' and m_2 to satisfy

$$m_1 z_2'/z_1' = m_2 z_2/z_1 \quad \text{and} \quad m(1 - z_2' / z_1') = m_2(1 - z_2/z_1); \qquad (1.5)$$

clearly this can be done since we have two equations and two unknowns, and the values will hold good for all points P_2. Thus we have shown that ideal image formation for two pairs of conjugate planes implies ideal imagery for all other pairs.

1.3 Properties of an ideal system

We can now develop many properties of ideal systems which are to be used in the paraxial approximation. For this purpose, we indicate the optical system schematically as in Fig. 1.3, but it must be understood that it may extend for a considerable distance along the axis and the constructions to be explained may take place inside the physical system, i.e. with virtual parts of rays.

In Fig. 1.3, let r_1 be a ray from the point at infinity on the axis in object space, i.e. a ray parallel to the axis. It meets the axis in image space at some point F′; this must be the image of the axial point at infinity in object space, since two rays pass through both these points, namely the ray r_1 and the ray

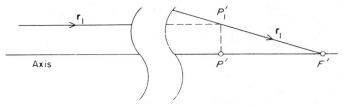

FIG. 1.3. Principal focus and principal point

along the axis, and we are assuming ideal image formation.† The point F′ is
called the second principal focus or image-side principal focus.

Let the segment of ray r_1 in object space be produced until it meets the
segment in image space at P'_1; a plane normal to the axis through P'_1 meets the
axis at P′, the second or image-side principal point, the plane itself being the
second or image-side principal plane.

Similar constructions and definitions lead to the object-side principal focus
and principal point. These constructions can be made on the same diagram
with the rays r_1 and r_2 parallel to the axis chosen to be at equal distances from

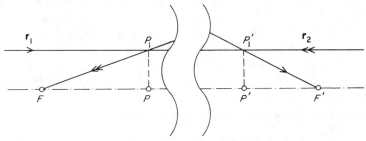

FIG. 1.4. Unity magnification between the principal points

the axis, as in Fig. 1.4, which shows all four points F, F′, P, P′. The two segments of r_1 meet at P'_1 and those of r_2 at P_1. Both rays r_1 and r_2 pass through
P_1 and P'_1, so these points must be object and image; furthermore, using the
properties of ideal image formation, the planes normal to the axis through P_1
and P'_1 must be conjugates and since by construction $PP_1 = P'P'_1$ the magnification between these planes must be unity. For this reason, P and P′ are
sometimes called unit points.

By their definitions, F and P are always in object space and F′ and P′ are
always in image space; however, it may very well happen, for example, that
F′ and P′ may physically lie to the left of the system, although they are still

† We defer until Section 3.4 the special case in which the ray r_1 emerges parallel to the axis.

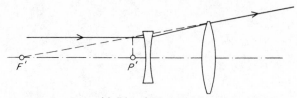

FIG. 1.5. An optical system with P' and F' physically on the left of the system

in image space. Fig. 1.5 shows a system consisting of two components in which this occurs.

The distance P'F' is the second or image-side focal length, denoted by f', and $PF = f$ is the object-side focal length. These magnitudes take signs according to the order of the letters in the definition in relation to the positive z direction, so that in Fig. 1.4 f would be negative and f' positive.

The four points F, F', P, P', along the axis fix the properties of the ideal optical system completely. We can use them in a construction to find the position and size of the image of any object, as in Fig. 1.6 where the optical system is represented merely by these four points and the principal planes. Let the object be OO_1 and let the ray r_1 be drawn through O_1 parallel to the axis to meet the first principal plane in P_1; it must emerge through P_1' at the same distance from the axis, and pass through F'. In the same way, the ray r_2 from O_1 through F is drawn through P_2 and P_2'. The point O_1' in which the image side segments of r_1 and r_2 meet must be the image of O_1, and O' must therefore be on the perpendicular from O_1' to the axis. This construction will be recognized as a simple generalization of the elementary construction for image formation by a thin lens.

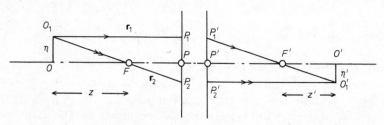

FIG. 1.6. Geometrical construction for conjugates

Figure 1.6 also yields simple formulae relating the positions and sizes of the object and image. Let η and η' be the object and image heights; these take signs according to the y-axis in the coordinate system (Fig. 1.1), so that in Fig. 1.6 η is positive and η' negative. Let $FO = z$, $F'O' = z'$, so that these quantities specify the axial conjugate positions; they are taken as directed segments with signs according to the z-axis, so that in Fig. 1.6 z is negative and z' is positive. From the similar triangles FOO_1 and FPP_2 we have

$$\eta'/\eta = -f/z \tag{1.6}$$

and likewise from $F'O'O_1'$ and $F'P'P_1'$,

$$\eta'/\eta = -z'/f'. \tag{1.7}$$

Combining eqns (1.6) and (1.7), we have

$$zz' = ff'; \tag{1.8}$$

this is known as a conjugate distance equation, since it relates z and z'; it is generally called Newton's conjugate distance equation. Isaac Newton gave it for a single surface ("Opticks", Book 1, Part 1, axiom 6, Dover 1952, based on the 4th edition 1730).

At the same time, we have obtained important expressions for the magnification $m = \eta'/\eta$ in eqns (1.6) and (1.7); these are generally written

$$z = -f/m, \qquad z' = -mf'. \tag{1.9}$$

It is also useful to have a conjugate distance equation in terms of the distances of object and image from the respective principal planes. Let $PO = l$, $P'O' = l'$, again with signs implied by the fact that PO is a directed segment; thus l is negative and l' positive in Fig. 1.6. We have

$$l = z + f, \qquad l' = z' + f'; \tag{1.10}$$

if the values of z and z' are substituted from eqn (1.9) and m is eliminated, we obtain

$$\frac{f'}{l'} + \frac{f}{l} = 1, \tag{1.11}$$

the required equation. We also have, analogous to eqn (1.9),

$$l = f\left(1 - \frac{1}{m}\right), \qquad l' = f'(1 - m) \tag{1.12}$$

and so

$$m = -\frac{l'f}{lf'}. \tag{1.13}$$

The considerable difference in form between the two conjugate distance equations, eqns (1.8) and (1.11), is because in eqn (1.11) the conjugates are referred to origins in object and image space which are themselves conjugates, namely the two principal points, whereas the principal foci are *not* conjugates. This brings to light a slight inconsistency in notation; primed and unprimed letters normally refer to conjugates or to some other quantities, e.g. angles of incidence and refraction, which have the relationship "before and after going through the optical system or one surface of it". The principal foci do not fit into this scheme and they are therefore occasionally labelled F_1 and F_2, but F and F' is the more usual notation.

A third useful pair of points on the axis can be defined, the nodal points N and N'; these are such that a ray entering through N emerges from N' parallel to its initial direction. We can find the positions of the nodal points by starting with the usual skeleton system specified by F, F', P, P' as in Fig. 1.7; we construct any ray r_1 through F, meeting the principal planes in P_1

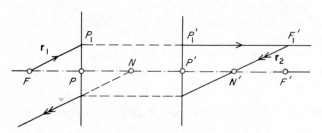

FIG. 1.7. Construction for the nodal points

and P_1' and meeting the image-side focal plane at F_1'; next we draw the ray r_2 through F_1' in image space and make it parallel to the segment of r_1 in object space. Since r_1 and r_2 meet at an image point F_1' which is on the image-side focal plane, they must come from an object point at infinity, i.e. they are parallel in object space: thus, both segments of r_2 are parallel to the segment of r_1 in object space and so r_2 must intersect the axis in the nodal points. It is easily seen by similar triangles that

$$FN = f', \qquad F'N' = f. \tag{1.14}$$

The six points F, F', P, P', N, N', are sometimes called the cardinal points. If either of the quadruples F, F', P, P', or F, F', N, N', is known the properties of the system are determined completely, since the conjugate distance equation and the magnification formulae are known. The points can occur in any order and relative positions on the optical axis, subject only to the restrictions implied by eqn (1.14).

The relation between axial object and image points given by eqns (1.8) and (1.11) is in effect a one-to-one correspondence between pairs of points on a line, the optical axis; it is an *involution*, in the terminology of projective geometry. Similarly, the transformation which expresses the image segment of a ray in terms of the object segment is a one-to-one correspondence between lines in the same three-dimensional space (a *collineation*) with axial symmetry.

The more detailed theory of involutions and collineations is not important in geometrical optics, but they are mentioned here to establish the point that the most general one-to-one correspondences take this form; on the other hand, it will be seen that in *real* optical image formation the relationship between object and image entities is more complex. For example, more than one axial "image point" may correspond to a single object point, on account of spherical aberration, and a point which is common to three rays in object space may not be common to the corresponding three rays in image space. Thus real optical image formation is essentially more complicated than the ideal case we have been discussing.

2 Geometrical Optics

2.1 Rays and geometrical wavefronts

We obtained in Chapter 1 a simple model of image formation with axial symmetry, and we pointed out there that this was based on assumptions about optical systems which are only valid under certain restrictions. In this chapter and the next we explain these restrictions and develop further the theory of optical systems within them. In order to do this, we have to introduce a further concept, the *geometrical wavefront*, in addition to the *ray*, already used.

The concept of a geometrical wavefront appears in the work of Fermat (1667), Malus (1808), Hamilton (1820–30) and others, as a surface of constant optical path from the source or a surface orthogonal to the rays from a source point. More recently the shape of the geometrical wavefronts has been used to characterize the aberrations of an optical system directly, rather than regarding the ray patterns as fundamental; one of the earliest authors to do this was G. Yvon ("Contrôle des surfaces optiques", Paris 1926). This usage of the geometrical wavefronts provides a link with the physical concepts which originated with C. Huygens (1690) and A. Fresnel (1866) and developed into the Kirchhoff diffraction theory (1891). A very full treatment of the early history of these topics is given by E. T. Whittaker (1951), "History of the Theories of Aether and Electricity", Vol. I, revised edition, Nelson, London.

To an adequate approximation, we can regard rays as the paths along which the radiation energy travels; this breaks down near foci and near the edges of shadows, owing to diffraction effects, but it is essential to geometrical optics that these are ignored. Now, let a point source of light be placed in

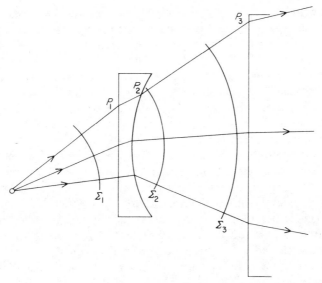

FIG. 2.1. Geometrical wavefronts and rays

front of any optical system, not necessarily axially symmetric, and suppose that the rays from this source O are traced through the system, as in Fig. 2.1, using the law of refraction; the details of the law of refraction will be discussed later (Section 2.2), but it is sufficient to know that we can calculate the ray paths according to simple rules. Recalling that light is propagated with a finite velocity, we can mark off on these rays sets of points which the light disturbance reaches in given times and we can join up the sets to form surfaces Σ_1, Σ_2 and Σ_3 in the figure, for example. We do this by noting that the velocity of light in a medium of refractive index n is c/n,† and so light travels from P_2 to P_3, say, in a time nP_2P_3/c, where n is the refractive index of the medium between P_2 and P_3; more generally, light travels from A to B in the time

$$t_{AB} = \frac{1}{c} \int_A^B n \, ds, \qquad (2.1)$$

where the integration is taken along the ray path and ds is an element of this path; thus the surfaces in Fig. 2.1 are simply surfaces of constant $\int n \, ds$ from the source point O. In the diagram, it is implied that the refractive index is constant along a ray segment and then changes discontinuously at a refracting surface; in this case the integral becomes simply a summation.

The quantity $\int n \, ds$ (or $\sum n \, \Delta s$ for most systems) is called the optical path

† For many purposes, it is convenient to regard this as a definition of refractive index.

length. The surfaces $\Sigma_1, \Sigma_2, \Sigma_3, \ldots$ are called *geometrical wavefronts* or simply *wavefronts* for brevity; they correspond approximately to surfaces of constant phase or wave epoch in the wave model of light, i.e. phasefronts. The approximation is good at large distances from foci or edges of shadow regions but near such regions the geometrical wavefronts do not correspond at all to true phasefronts. The most familiar example is a single mode TEM_{00} or Gaussian laser beam, where at the focus, or beam waist as it is usually called, the phasefront is plane but the geometrical wavefront contracts to a single point; see Appendix D for a discussion of the propagation of Gaussian beams. The approximation may also be poor when the geometrical wavefronts are almost plane, as explained in Section 13.3, but in most practical cases geometrical wavefronts and phasefronts can be taken to coincide. In geometrical optics the wavefronts are surfaces of constant optical path length from the source point O. In a system such as in Fig. 2.1 we can imagine a double infinity of rays from O and a single infinity of wavefronts; this ensemble is sometimes referred to as a pencil and it is often convenient in discussing aberrations to select various chosen rays and wavefronts from a hypothetical complete pencil.

2.2   Snell's law of refraction

The wavefronts as defined in the preceding section are, with rays, the basic concepts of geometrical optics, and optical path length is the basic physical quantity. We next have to show how the rays and wavefronts are propagated through lens surfaces, i.e. we have to investigate refraction.

In terms of rays we can express refraction by Snell's law; let I and I' be the angles of incidence and refraction at an interface between media of refractive indices n and n', as in Fig. 2.2; then Snell's law states that the incident and refracted rays are coplanar with the normal and that

$$n' \sin I' = n \sin I. \tag{2.2}$$

Snell's law can be put more compactly in vector form; if r and r' are unit vectors along the incident and refracted rays and n is a unit vector along the normal to the interface, then

$$n'(r' \wedge n) = n(r \wedge n), \tag{2.3}$$

for the absolute magnitudes of the two sides of this equation are equal to the two sides of eqn (2.2), and the vector equality ensures the coplanarity of the rays and the normal. This form of the equation will be needed for tracing rays in three dimensions.

According to E. T. Whittaker ("History of the Theories of Aether and

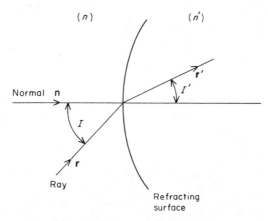

FIG. 2.2. Snell's law of refraction

Electricity", Vol. 1, 1951), W. Snell (1591–1676) discovered the law experimentally about 1621; he did not publish it but communicated it privately to several people including R. Descartes. The latter published it in 1637.

Snell's law can be verified experimentally by means of accurate goniometers to a fraction of a second of arc. It also follows from the electromagnetic wave theory of light, so we can regard it as given.

The quantities n and n' appear now merely as physical constants to be determined for the media on either side of the interface and in fact refractive index is almost invariably determined for optical instrumentation purposes by methods based very directly on Snell's law. The proof that this quantity is the same as that in the formula c/n for the speed of light in a material medium again appears in the derivation of Snell's law from electromagnetic theory. It should be remarked also that the refractive index varies with wavelength or frequency for all material media; this is called *dispersion* and it is the source of chromatic aberrations in optical systems.

It is usual to include reflection as a special case in the formulation of Snell's law by adopting the convention that n' is replaced by $-n$ when the ray is incident as in Fig. 2.2 and the interface is reflecting. This formal device yields the correct result from the vectorial form, provided we accept the following convention about ray directions. The directions of all rays are referred to a right-handed coordinate system as in Fig. 2.3, and the direction cosines of the line along the ray are used to specify its direction; these are, of course, the same as the components of a unit vector which starts from the origin, is parallel to the ray and has its termination to the right of the x–y plane. For example, suppose we have a ray which passes through the origin and which also lies in the octant with positive x, y and z, as in Fig. 2.3; then the corre-

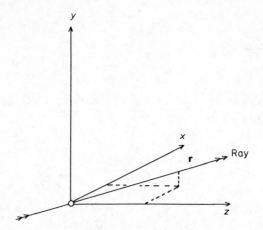

FIG. 2.3. Signs of ray components; the coordinate system is right handed, so that the x-axis points into the paper

sponding unit vector has its three components all positive, and the ray direction is denoted in our convention by positive components irrespective of the direction of travel of the light along it. If the light were travelling in the reverse direction from that indicated by the double arrow, then this fact would be indicated by there being a negative refractive index in the medium. With this convention, eqns (2.2) and (2.3) include reflection when n' is taken as $-n$. This is illustrated in Fig. 2.4, where r'' is the unit vector along the reflected ray direction as given by this convention and $-nr''$ is the actual reflected ray; thus

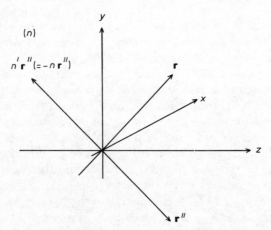

FIG. 2.4. Directions and signs of reflected rays

the geometrical line representing the reflected ray is correct and the factor $n'' = -n$ gives the reversal.

If $n' < n$ in eqn (2.2) when $\sin I = n'/n$ we find $\sin I' = 1$ and for greater values of $\sin I$ the value of I' would be purely imaginary. In fact there is no refracted ray when this happens and the light is completely reflected, an effect known as *total internal reflection*. This has many applications in the design of prisms.

2.3 Fermat's principle

An entirely different approach to refraction is provided by Fermat's principle, which is a stationarity principle of the same kind as the principle of least action of Maupertuis. It was originally proposed by Pierre de Fermat in 1657 as a principle of least time: the path taken by light travelling from A to B through any optical system will be such that the time of travel is a minimum. This can easily be put in terms of optical path lengths by means of eqn (2.1): of all possible paths (straight line segments or otherwise) between A and B, that which has the shortest optical path length represents a physically possible ray.

As stated in this form, the principle only applied to points A and B which are close enough to ensure that there is no real focus between them†; if A and B are general points, then the optical path length along the physically possible ray is not necessarily a minimum, but it is *stationary*. The most general explanation of the term *stationary* involves the calculus of variations, and a good exposition is given by M. Born and E. Wolf ("Principles of Optics", 3rd edition 1963, Pergamon Press, London); for the present purposes, it may be sufficiently explained in terms of paths from A to B consisting of straight line segments, as in Fig. 2.5, where the full line indicates the physically possible path and the broken line some other path. The optical path

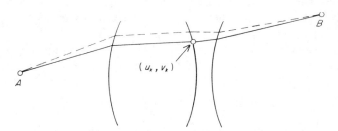

FIG. 2.5. A stationary optical path

† This is a loose statement of the restriction.

length W from A to B is then a function of pairs of parameters u_k, v_k, ($k = 1, 2, \ldots$) which are generalized coordinates of the points where the line segments meet the refracting surface. Then stationarity of the optical path means that for the physically possible ray these coordinates must be such that

$$\frac{\partial W}{\partial u_k} = \frac{\partial W}{\partial v_k} = 0 \quad (k = 1, 2, \ldots). \qquad (2.4)$$

It should be emphasized that in this discussion A and B are not restricted to be in any sense object and image—they are general points.

The straight line propagation of light, which we have already tacitly assumed, is an immediate consequence of Fermat's principle. It is also easy to obtain Snell's law, as in Fig. 2.6; let the y–z plane be the refracting surface,

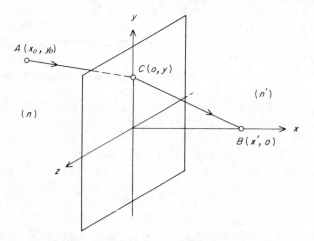

FIG. 2.6. Derivation of Snell's law from Fermat's principle

separating media of index n and n', and let the points A and B have coordinates (x_0, y_0) and $(x', 0)$. There is no loss of generality in taking a plane refracting surface, since refraction according to either Snell's law or Fermat's principle depends only on the local direction of the tangent plane at the point of incidence C; also it is clear that C lies in the x–y plane by symmetry.

The optical path W from A to B is then

$$W = n\{x_0^2 + (y - y_0)^2\}^{\frac{1}{2}} + n'\{x'^2 + y^2\}^{\frac{1}{2}}. \qquad (2.5)$$

The condition of stationarity with respect to changes in y is

$$\frac{\partial W}{\partial y} \equiv \frac{n(y-y_0)}{\{x_0^2 + (y-y_0)^2\}^{\frac{1}{2}}} + \frac{n'y}{\{x'^2 + y^2\}^{\frac{1}{2}}} = 0; \qquad (2.6)$$

the coefficients of n and n' can be identified as the sines of the angles of incidence and refraction, with suitable sign convention, and thus Snell's law is obtained.

A direct verification of Fermat's principle by experiment is not, of course, possible but, as in the case of Snell's law, it is possible to deduce Fermat's principle from electromagnetic theory provided it is assumed that regions of the electromagnetic field remote from foci, shadow edges and other places where strong diffraction effects might occur are excluded; the whole concept of rays breaks down in such regions. Treatments of this kind are given, for example, by Born and Wolf ("Principles of Optics", 3rd edition 1963, Pergamon Press, Oxford), by A. Sommerfeld ("Optics", English translation, O. Laporte and P. A. Moldauer, Academic Press, New York, 1954) and by R. K. Luneburg ("Mathematical Theory of Optics", Wiley–Interscience, New York, 1965); all these authors adopt different approaches which may be regarded loosely as treatments of Maxwell's equations of the electromagnetic field in which the wavelength of the light is allowed to tend to zero. In this way all the usually accepted attributes of rays and geometrical wavefronts can be obtained. A comprehensive survey of this question of the transition from electromagnetic theory to geometrical optics with a discussion of the limitations of each approach is given by M. Kline and I. W. Kay, "Electromagnetic Theory and Geometrical Optics" (Wiley–Interscience, New York, 1965).

2.4 The laws of geometrical optics

For our purposes it is convenient to regard both Snell's law and Fermat's principle as basic postulates of geometrical optics, although, as we have pointed out, they are both consequences of electromagnetic theory. It is convenient also to mention explicitly the other properties of rays and wavefronts in geometrical optics. These are rectilinear propagation in homogeneous media, reversibility of ray paths, non-interference of intersecting ray paths and the inverse square law of illumination for a point source.

We can immediately derive from Fermat's principle the important result that rays are normals to wavefronts,[†] the theorem of Malus and Dupin. Let Σ and Σ' in Fig. 2.7 be two wavefronts of a pencil from a point source; the wavefronts are on either side of the kth refracting surface S of an optical system between media of refractive indices n and n'; let PQP' and $P_1Q_1P_1'$ be two rays and suppose that the theorem is true up to the medium of index n. We have

$$[PQP'] = [P_1Q_1P_1'], \qquad (2.7)$$

[†] It is assumed here and everywhere else in this book that we are dealing only with isotropic media.

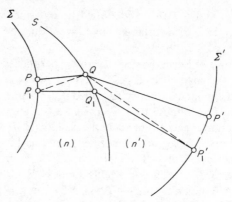

FIG. 2.7. The theorem of Malus and Dupin

where the square brackets denote optical path lengths, by the definition of wavefronts. The path P_1QP_1', shown in broken line in the figure, is not a true ray path, but, by Fermat's principle

$$[P_1Q_1P_1'] = [P_1QP_1'] + O(\varepsilon^2) \tag{2.8}$$

where $\varepsilon = QQ_1$ and the O( ) symbol has its usual significance. Thus

$$[P_1QP_1'] = [PQP'] + O(\varepsilon^2) \tag{2.9}$$

But, since PQ and P_1Q_1 are normals to Σ we have

$$[P_1Q] = [PQ] + O(\varepsilon^2), \tag{2.10}$$

since PP_1 is of the same order of magnitude as QQ_1, and thus, by subtracting eqn (2.10) from eqn (2.9), we have

$$[QP_1'] = [QP'] + O(\varepsilon^2), \tag{2.11}$$

or QP' is normal to Σ'. But, since the pencil originated at a point source, the wavefronts in the object space must have been spherical and normal to the rays; thus the theorem is proved by induction.

This theorem, that rays are normals to wavefronts, enables us conveniently to develop the analogue for real optical systems of the ideal image formation described in Chapter 1; this analogue is called Gaussian optics or paraxial optics, the latter name implying that it is valid only in a region close to the optical axis. The development is facilitated by the immediate deduction from the theorem of Malus and Dupin that, if the rays of a pencil intersect in a single point after refraction through an optical system, then the wavefronts will be spherical,† and vice versa. Thus we have two equivalent definitions of

† Here and elsewhere we use "spherical" to mean "portions of spheres".

the formation of a perfect point image, that all the image-forming rays meet in a single point or that the wavefronts are spherical. Departures from these conditions imply the presence of *aberrations* and these will be defined in more detail in Chapter 6.

3 Gaussian Optics

3.1 The domain of Gaussian optics

If a point source is on the axis of a symmetrical optical system, then the wave-fronts of the pencil in image space must, by symmetry, be figures of revolution about the axis. Taking a coordinate system as in Fig. 1.1, i.e. with the z-axis along the axis of symmetry, we can write the equation of such a figure of revolution in the form

$$z = \tfrac{1}{2}c(x^2 + y^2) + \mathrm{O}((x^2 + y^2)^2), \tag{3.1}$$

assuming the origin to be chosen to lie on the surface.† In the same way, any refracting surface can be represented by a similar equation; for example, a sphere of curvature c would be represented by

$$z = \tfrac{1}{2}c(x^2 + y^2) + \tfrac{1}{8}c^3(x^2 + y^2)^2 + \mathrm{O}((x^2 + y^2)^3). \tag{3.2}$$

This power series development immediately suggests that we should examine the approximation obtained by ignoring all terms but the first, quadratic, term. This was first done in complete generality by K. F. Gauss in a celebrated memoir, "Dioptrische Untersuchungen", published in Göttingen in 1841, although others, notably Roger Cotes over a century earlier, had made considerable progress along similar lines.

We thus define the domain of paraxial or Gaussian optics to be close enough to the axis to ensure that all terms of higher order of magnitude than quadratic in x and y are to be neglected. This definition has the immediate

† It is also a tacit assumption that the surface does not have a cusp or other singularity on the axis.

consequence that there is perfect image formation for object points on the axis if we restrict the rays to the Gaussian region, since the equation of any refracted wavefront must, to this approximation, be of the form

$$z = \tfrac{1}{2}c(x^2 + y^2) \tag{3.3}$$

and this cannot be distinguished from a sphere of curvature c. In a similar way, we can state that all reflecting or refracting surfaces are spherical, in Gaussian approximation, since, by eqn (3.2), other surfaces of the same axial curvature such as paraboloids or ellipsoids of revolution differ from a sphere only in the terms higher than quadratic.

We next have to define the extent of the Gaussian region for object points which are not on the axis. Let O and O' be axial conjugates in Fig. 3.1, and

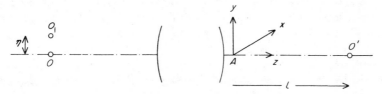

FIG. 3.1. Off-axis image formation

let O_1 be an off-axis object point at a distance η from O in a plane perpendicular to the optical axis. We wish to choose a restriction on the order of magnitude of η which will similarly ensure perfect image formation for O_1. Let the equation of a wavefront of the pencil converging to O' be

$$z = \frac{1}{2l}(x^2 + y^2), \tag{3.4}$$

where the origin is chosen arbitrarily at A and l is the distance AO'. The pencil from O_1 will travel through this region after refraction, and we can pick out the wavefront which passes through A; let its equation be

$$z = \frac{1}{2l}(x^2 + y^2) + F(x, y, \eta) \tag{3.5}$$

where the function F is to be determined. If we assume again that this wavefront has not a singularity at A then F can be written as a power series giving

$$z = \frac{1}{2l}(x^2 + y^2) + \eta(ax + by + c)$$
$$+ \eta^2(dx^2 + exy + fy^2 + gx + hy + j) + O(\eta^3). \tag{3.6}$$

Now in the right-hand side of this equation we can set c and j equal to zero, since they do not contain x or y and we have postulated that the wavefront passes through the origin A. Also the whole of Fig. 3.1 is symmetrical about the y–z plane, since we assume O_1 to lie in this plane, so that eqn (3.6) must contain only even terms in x, i.e. we set a, e and g equal to zero. It can be seen by inspection of the remaining terms that, if only terms of order ηx and ηy are retained, the resulting wavefront is still spherical, it has the same curvature as the axial wavefront and its centre is displaced from O' in a direction perpendicular to the axis; the equation under this restriction is

$$z = \frac{1}{2l}(x^2 + y^2) + b\eta y, \tag{3.7}$$

which, in our approximation, represents a sphere of radius l and with centre at a distance $-lb\eta$ above O'. Clearly, this distance is proportional to η for a given system and a given position of O; thus we have found restrictions under which all object points have perfect point images and objects in a plane perpendicular to the axis form geometrically similar images also in a plane perpendicular to the axis.

The restrictions may be re-stated compactly in the form: terms in the wavefront shape depending on higher powers than the square of the aperture and object coordinates are to be neglected. The paraxial or Gaussian region is so defined, and if we apply the reasoning of Section 1.2 to the result of the above paragraph we see that within the Gaussian region there is ideal image formation with all the properties obtained in Section 1.3; the definitions of foci, principal planes, focal lengths, etc., and the constructions and equations for conjugates are all immediately applicable.

Thus, any axisymmetric optical system has ideal image formation in the sense of Chapter 1 in the Gaussian region; however, at distances from the axis at which the approximations of Gaussian optics are invalid this is no longer true and the system will have aberrations. It seems to be true that no optical system (with certain trivial exceptions such as plane mirrors) can have ideal image formation in a finite region beyond the Gaussian region and only certain restricted kinds of freedom from aberrations are possible. Optical designers are therefore concerned with reducing aberrations to chosen tolerance levels rather than to zero.

The choice of the particular powers of the aperture and object size to be included as defining the Gaussian region may appear somewhat arbitrary at this stage; in Chapter 7 it will be shown that the complete properties of a symmetrical optical system can be expressed by a power series in three variables, representing coordinates in the aperture and the object size or field, and that the lowest degree terms represent paraxial image formation while the higher terms correspond to aberrations; in this way it will appear that the definition of the paraxial region is natural and inevitable.

The above definition of the Gaussian region in terms of the degrees of terms in a power series which are to be neglected can be amplified by considering the physical magnitudes of the terms neglected. Lord Rayleigh suggested in 1880 that a pencil could be regarded as substantially aberration-free if the wavefront were within a quarter of a wavelength of a true sphere, since then the light disturbances arriving at the focus would all reinforce each other in phase; this is the so-called Rayleigh quarter-wavelength rule. Thus, we may say that physically the higher order terms which are neglected in the Gaussian approximation ought to amount to less than a quarter of a wavelength, or, in other words, the Gaussian region is the region so close to the axis that within it all deviations of wavefronts and refracting surfaces from true spherical shape are less than a quarter wavelength. It is found that on the basis of this definition the Gaussian region has quite a useful extent and it is not merely a limiting concept.

3.2 Definitions; the relationship between the two focal lengths

As we remarked in 3.1, all the definitions and properties of ideal image formation obtained in 1.3 apply immediately to Gaussian optics; for completeness we recapitulate these with somewhat different emphasis.

The image side principal focus F' is the image of an axial object point at infinity, and similarly for the object side principal focus F. The object and image side principal points P and P' are the axial points between which there is unit transverse magnification; the plane perpendicular to the axis through P', i.e. the image side principal plane, is obtained as the locus of intersections of the object and image space segments of the rays used to define F', as in Fig. 1.3, and similarly for the object side principal plane. Strictly these planes are now defined only very near the axis, in the Gaussian region, but we shall see in Section 3.10 that they can be extended to a greater distance with a suitable convention. The distances $PF = f$ and $P'F' = f'$ are the two focal lengths, with signs defined by the order of the letters and the sign convention for the z-axis.

The relation between axial conjugates O and O' is given in terms of the distances $FO = z$ and $F'O' = z'$ by the Newton conjugate distance equation

$$zz' = ff'; \qquad (3.8)$$

alternatively in terms of $PO = l$ and $P'O' = l'$ we have

$$\frac{f'}{l'} + \frac{f}{l} = 1. \qquad (3.9)$$

We have also the transverse magnification equations (eqns (1.9) and (1.12)) and the definitions of the nodal points N and N'.

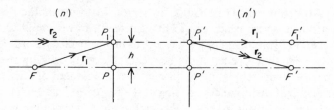

FIG. 3.2. The ratio between the two focal lengths

The first new result which we can obtain by the application of geometrical optics is a relation between the two focal lengths. In Fig. 3.2 let F, P, P′ and F′ be the principal foci and principal points of an optical system, and let a paraxial ray r_1 through F meet the principal planes at P_1 and P_1' and the image side focal plane at F_1'. We can regard $P'P_1'$ and $F'F_1'$ as wavefronts of the pencil from F and so we have

$$[FP_1P_1'F_1'] = [FPP'F'], \tag{3.10}$$

or

$$n.FP_1 + [P_1P_1'] + n'f' = -nf + [PP'] + n'f'. \tag{3.11}$$

Thus, if the distance PP_1 is denoted by h, we have on expanding FP_1 by the binomial theorem

$$\frac{-nh^2}{2f} = [PP'] - [P_1P_1']. \tag{3.12}$$

Similarly by taking another ray r_2 which enters the system parallel to the axis and at the same height h, we obtain

$$\frac{n'h^2}{2f'} = [PP'] - [P_1P_1'] \tag{3.13}$$

and comparing eqns (3.12) and (3.13) we obtain the required relationship:

$$\frac{n'}{f'} = \frac{-n}{f}. \tag{3.14}$$

From eqn (3.14) it is seen that if the refractive indices of the object space and the image space are equal, then the two focal lengths are equal in magnitude and opposite in sign. This is, of course, the most frequent situation. The quantity n'/f' $(= -n/f)$ is called the *power* of the optical system and it is denoted by the symbol K; the power is frequently used as a measure of strength of an optical system because, as will be seen, it occurs naturally in many equations involving raytracing and the combination of optical systems; also, its value is unchanged if the system, with terminating media, is reversed end to end. The reciprocal of K, which has the dimension of length, is called the equivalent focal length, usually abbreviated to efl. It is in fact the image side focal length which a system with air† on both sides would have if it had the same power as the system in question. Thus for a system in air the image side focal length is the efl, and it is the reciprocal of the power. It is easily shown that, in terms of the power, eqn (3.9) takes the form

$$\frac{n'}{l'} - \frac{n}{l} = K. \tag{3.15}$$

3.3 The Lagrange invariant and the transverse magnification

In Fig. 3.3 a pair of axial conjugates O, O′ is indicated and also the off-axis pair O_1, O_1'; let OO_1 and $O'O_1'$ be denoted by η and η' respectively, so that the transverse magnification m is η'/η. From eqn (1.13) we have

$$\frac{\eta'}{\eta} = \frac{-l'f}{lf'} \tag{3.16}$$

and so from eqn (3.14)

$$\frac{\eta'}{\eta} = \frac{nl'}{n'l} \tag{3.17}$$

or

$$\frac{n'\eta'}{l'} = \frac{n\eta}{l}. \tag{3.18}$$

Now let h be the height at which the imaging ray shown in Fig. 3.3 meets the principal planes, ‡ and let u and u' be defined by

$$u = -h/l, \qquad u' = -h/l'; \tag{3.19}$$

if we multiply eqn (3.18) through by h we obtain

$$n'u'\eta' = nu\eta. \tag{3.20}$$

The quantities u and u' are defined only in the Gaussian region, but in this region they clearly represent angles and they are called convergence angles. Thus eqn (3.20), when written in the form

$$\frac{\eta'}{\eta} = \frac{nu}{n'u'} \tag{3.21}$$

gives the transverse magnification in terms of the convergence angles of the

† By "air" we mean strictly a medium of unit refractive index, i.e. vacuum, but the usage is general in geometrical optics.

‡ It meets them both at the same height because they are planes of unit magnification.

ray from the axial object point. The sign of the convergence angle is given by its definition in terms of h and l in eqn (3.19), so that in Fig. 3.3 u is positive and u' is negative; this convention is then extended to other paraxial angles, and it agrees with the usual convention that an anti-clockwise rotation from the axis is positive.

Equation (3.20) has a much wider application than to the calculation of transverse magnification. Let us suppose that the optical system of Fig. 3.3

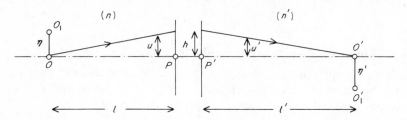

FIG. 3.3. Transverse magnification

is to be followed by a further system; then the quantity $n'u'\eta'$ for the first system will be the $nu\eta$ for the second, for the refractive index is the same, the image for the first system becomes the object for the second and the convergence angle is the same. Thus this quantity is an invariant through both systems and, therefore, it must be an invariant right from the object space through all intermediate spaces of any symmetrical optical system. This invariant was first given by Robert Smith in his "Compleat System of Opticks", Book II, published in 1738. It was later given in various forms by Lagrange and by Helmholtz. Following general usage we call it the Lagrange invariant and denote it by the symbol H.

The Lagrange invariant enters into aberration calculations, as will be seen in Chapter 8; it is significant in the photometry of optical instruments, since it can be shown that the total flux collected by a system from a uniformly radiating object is proportional to H^2; also the Lagrange invariant divided by the wavelength of the light is, in effect, the unit in the dimensionless coordinates used in the diffraction theory of optical instruments. On account of the importance of these varied applications, we now give a proof from first principles of eqn (3.20).

Let OO_1 and $O'O'_1$ be conjugates of heights η and η' (Fig. 3.4) and let the ray r_1 from O to O' have convergence angles u and u'. Rays r_2 and r_3 from O_1 to O'_1 are drawn respectively parallel to ray r_1 and to the axis in object space and O_1P and O'_1P' are perpendiculars from O_1 and O'_1 to ray r_1. Then

$$[OO']_{r_1} = [OO']_{axis}, \qquad (3.22)$$

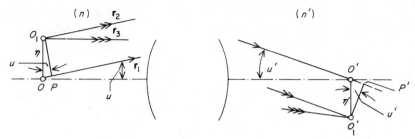

FIG. 3.4. The Lagrange invariant

since O and O' are conjugates, and similarly

$$[O_1O_1']_{r_2} = [O_1O_1']_{r_3}. \tag{3.23}$$

We can now take OO_1 to represent a wavefront of a pencil from the axial object at infinity, so that ray r_3 and the axis will be rays of this pencil; the wavefront of this pencil which passes through O' will be spherical and tangent to $O'O_1'$ at O' so that we have

$$[O_1O_1']_{r_3} = [OO']_{\text{axis}} + O(\eta^2). \tag{3.24}$$

Combining eqns (3.22), (3.23) and (3.24), we obtain

$$[O_1O_1']_{r_2} = [OO']_{r_1} + O(\eta^2), \tag{3.25}$$

or

$$[O_1O_1']_{r_2} = [OP] + [PP']_{r_1} + [P'O'] + O(\eta^2). \tag{3.26}$$

Again we can regard O_1P as a wavefront of a pencil of which r_1 and r_2 are rays and we have

$$[O_1O_1']_{r_2} = [PP']_{r_1} + O(\eta^2) \tag{3.27}$$

and subtracting from eqn (3.26) we have

$$[OP] = [O'P'] + O(\eta^2). \tag{3.28}$$

Thus to the approximation of Gaussian optics we obtain again

$$nu\eta = n'u'\eta'. \tag{3.20}$$

This proof is independent of previous results, except the general theorem that imagery in the Gaussian region is perfect, and it shows the fundamental nature of the Lagrange invariant.

It may happen that in one of the spaces of an optical system, not necessarily the object space or the image space, the intermediate object/image is at infinity; this is then said to be a star space.† The form of the Lagrange invariant given in eqn (3.20) becomes indeterminate in a star space, since u tends to zero and η tends to infinity, but the indeterminacy may be resolved by writing u as $-h/l$, where h is the incidence height at a surface bounding the space and l is the intersection length from this surface; also we put $\eta = \beta l$, where β is a *field angle* defined with the same sign convention as for convergence angles. It is then seen that as l tends to infinity we have

$$H \rightarrow -nh\beta. \tag{3.29}$$

Any ghost surface in the star space can be used to determine h and l in the above; for l tends to infinity, so the origin is unimportant, and the ray which gives h becomes parallel to the axis. The field angle β is defined here only for a star space; the more general concept is that of the convergence angle of the principal ray (Section 3.5) and it will be shown in Section 5.2 that this leads to a simple alternative derivation of eqn (3.29).

3.4 Afocal systems and star spaces

The definitions of the foci and the principal points given in Section 1.3 break down in the special case when the ray from the axial object point at infinity emerges parallel to the axis, as in Fig. 3.5. Such systems form an important special class; they are said to be afocal or telescopic and they have the property that the image space becomes a star space if the object space is chosen to be a star space. The imaging properties of an afocal system for infinite conjugates are obtained from the star space form of the Lagrange invariant, given in eqn (3.29). Let h and h' be the incidence heights of the ray r_1 from the axial object point and let a ray r_2 from an off-axis object point at infinity have field angles β and β' in object and image space; these quantities

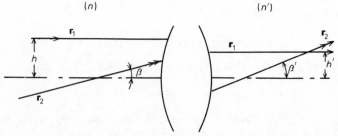

FIG. 3.5. An afocal system

---

† Being a star space is not, of course, an intrinsic property of a space in an optical system; any space can be a star space if the conjugate in it is chosen to be at infinity.

are indicated in Fig. 3.5. Then we have immediately

$$n'h'\beta' = nh\beta \tag{3.30}$$

so that the *angular magnification* is

$$\frac{\beta'}{\beta} = \frac{nh}{n'h'}. \tag{3.31}$$

If an afocal system is used with conjugates at a finite distance it can be seen from Fig. 3.5 that the transverse magnification is constant, for it is equal to h'/h. We cannot obtain conjugate distance relations analogous to eqn (1.8) or eqn (1.11) for an afocal system since there are neither foci nor principal planes to be used as origins for the measurement of conjugate distances.

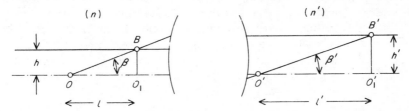

FIG. 3.6. Conjugate distance equation for an afocal system

However, we can choose arbitrarily a pair of axial conjugates as origins, such as O and O' in Fig. 3.6; we then construct another pair of conjugates O_1B and O'_1B' from the intersections of the rays through O and O' with a ray parallel to the axis. If then $OO_1 = l$, $O'O'_1 = l'$, we have from the figure

$$\frac{l'}{l} = \frac{h'\beta}{h\beta'}, \tag{3.32}$$

so that from eqn (3.31)

$$\frac{l'}{l} = \frac{n'h'^2}{nh^2}. \tag{3.33}$$

This is the required conjugate distance equation; it is clearly independent of the particular pair of conjugates O and O' chosen as origins, a fact which may be restated in slightly different form as, **the longitudinal magnification is constant and equal to the transverse magnification divided by the angular magnification.**

Figure 3.7 shows how an afocal system might be formed from two thin lenses of equivalent focal lengths f'_1 and f'_2; the lenses are placed so that the foci F'_1 and F_2 coincide and it may be seen that the transverse magnification

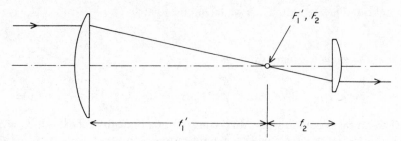

FIG. 3.7. Construction of an afocal system

is f_2/f_1' and, if the lenses are in air, the angular magnification is f_1'/f_2. This is, of course, essentially the arrangement of a refracting telescope and it is for this reason that afocal systems are also called telescopic.

Systems as in Fig. 3.7 are used as beam expanders for laser beams, the expansion or contraction ratio being simply the transverse magnification. The real intermediate image is useful since a small aperture or pinhole is often placed there to exclude unwanted stray light.

3.5 The aperture stop and the principal ray

If rays are traced from the axial object point through an optical system, the size of the pencil which can be accepted by the system will be limited by the size of one of the components, or by a *stop* or *diaphragm* deliberately introduced for this purpose. This limiting aperture is called the *aperture stop* and Fig. 3.8 shows a typical arrangement. If we imagine an observer at O, the

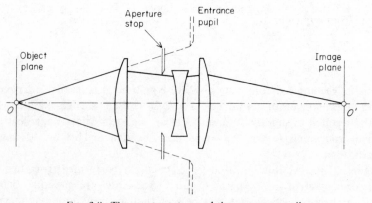

FIG. 3.8. The aperture stop and the entrance pupil

axial object point, he could in principle see all the rims of the lenses imaged in the part of the system to the left of each and it can be seen that the aperture stop will be that particular rim of which the image subtends the smallest angle at O. This image, indicated in the figure, is called the *entrance pupil*. In the same way, an observer at O′ could see images of all the rims formed by the parts of the system to the right of each and the image subtending the smallest angle at O′ is the *exit pupil*; the exit pupil is the image of the aperture stop, since this must be the limiting aperture from both sides of the system.

The aperture stop and the pupils have diameters that are usually well beyond the Gaussian region as defined in Section 3.1 and so strictly we must imagine the rays discussed here as non-paraxial or finite rays; in fact we shall see in Section 3.10 that the definitions and properties can for many purposes be validly expressed in terms of paraxial rays. Bearing this in mind we see that the pupils also limit the sizes of pencils from off-axis object points. Thus in Fig. 3.9 we have off-axis conjugates O_1 and O_1' and we can draw in the pencil

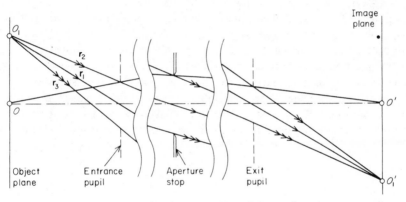

FIG. 3.9. Off-axis pencils and the pupils

from O_1 to O_1' as rays such as r_2 and r_3 from O_1 to the rim of the entrance pupil. The ray r_1 from O_1 to the centre of the pupil is in a sense the central ray of this pencil; it is called the *principal ray* (sometimes chief ray or reference ray); it also passes through the centres of the aperture stop and the exit pupil.

The pupils as seen from O_1 or O_1' will appear elliptical and in addition some portions may be cut off by lens rims which do not encroach on the axial pencil, an effect called *vignetting*. This effect is illustrated in Fig. 3.10. The

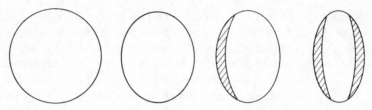

FIG. 3.10. Off-axis pupils and vignetting

choice of the ray of an off-axis pencil to be regarded as the principal ray is somewhat arbitrary under such conditions, particularly as the pupil imagery, i.e. the image formation from the entrance pupil to the exit pupil, may have large aberrations; however, as suggested above, the essential points involve only Gaussian considerations.

The aperture stop is usually physically inside an optical system; quite frequently the pupils are near the principal planes, but this need not be the case. The choice of the position of the stop depends sometimes in part on mechanical considerations, as when it is incorporated as an iris diaphragm in a photographic objective; however, the diameter and axial position of the stop also affect the aberrations of the system to the extent that, as will be shown in Chapter 8, an axial shift of the stop changes the relative proportions of different aberrations. Another factor which sometimes affects the choice of the position of the aperture stop is the inclination of the principal rays to the axis in object space or image space. It may happen that it is desirable that the principal rays in image space should all be parallel to the axis; this would involve an exit pupil at infinity and this is achieved by putting the aperture stop at the object side principal focus, as in Fig. 3.11; the aperture stop is then said

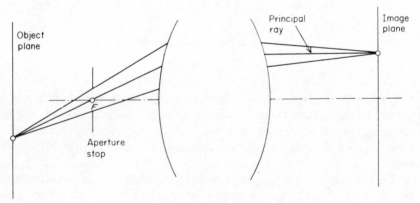

FIG. 3.11. Telecentric aperture stop

to be telecentric. Of several possible reasons for requiring this we mention the following two: (a) a measuring graticule may have to be put at the image plane and then small errors of focus will have no effect on the size of the image as measured on the graticule; (b) if a photograph is to be taken on film then slight departures from flatness of the film will not produce apparent distortions of image size and shape, an application of importance in the photography of bubble chambers.

If a system is afocal it may happen that rays from infinity on the axis in object space cross the axis in some intermediate space; then if the aperture stop is placed at that position the pupils will be at infinity in both object and image spaces, i.e. the system is telecentric on both sides.

3.6 Field stops

There may also be a stop in an optical system which limits the size of the field and this, if it exists, will be at the object or image plane or some intermediate real conjugate. For example, in a camera the field stop is a rectangular diaphragm in contact with the film and in a slide projector it is a similar diaphragm in contact with the slide; certain types of eyepiece usually have a field stop at the position of a real image inside the eyepiece.

Field stops do not have the basic optical importance that the aperture stop has since they do not control the aberration balance; this is, of course, because the field stop can only be placed at the object conjugate in each space. The field stop limits the field of view to that for which the optical system has adequate aberration correction and it also provides a format of the required shape. However, some optical systems do not have a definite field stop, for example the human eye. The images of the field stop in object and image space are sometimes called the entrance and exit windows, but little importance attaches to this concept.

Other stops found in optical systems are for trapping unwanted light from reflections and scatter off optical and non-optical surfaces and for preventing light from travelling along inappropriate paths in reflecting systems.

3.7 Gaussian properties of a single surface

In order to carry the subject further we now have to consider how to find the Gaussian properties of actual systems. Consider a single refracting (or reflecting) surface of curvature c which separates two media of indices n and n', as in Fig. 3.12. Clearly the principal planes coincide at the surface, since an object at the surface will have its image there and of the same size; we are then justified in measuring conjugate distances l and l' from the surface, as

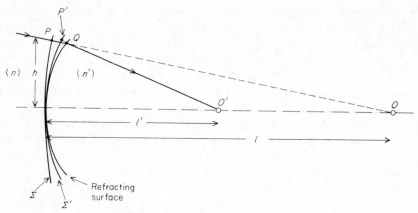

FIG. 3.12. A single refracting surface

indicated. Let Σ and Σ' be two wavefronts of the axial pencil before and after refraction, respectively, which both touch the refracting surface on the axis; in our figure Σ' happens to be a virtual wavefront, but this is of no significance. Let the image-forming ray shown meet the wavefronts at P and P' and the refracting surface at Q. Then since the distances between two wavefronts of the same pencil along any two rays are equal we have

$$[PQP'] = 0 \tag{3.34}$$

or,

$$n'.P'Q = n.PQ. \tag{3.35}$$

This equation is correct without any restriction to the Gaussian region, but within this region we have

$$PQ = \tfrac{1}{2}h^2 c - \tfrac{1}{2}h^2/l, \tag{3.36}$$

where h is the incidence height of the ray, and similarly for P'Q. Substituting in eqn (3.34) we obtain on cancelling the factor h^2,

$$n'\left(c - \frac{1}{l'}\right) = n\left(c - \frac{1}{l}\right). \tag{3.37}$$

Rearranging this result in the form

$$\frac{n'}{l'} - \frac{n}{l} = (n' - n)c \tag{3.38}$$

we recognize it as a conjugate distance equation of the form of eqn (3.15) and we identify the quantity $(n' - n)c$ as the power of the surface. Thus the two focal lengths are respectively

$$f = \frac{-n}{(n' - n)c}, \qquad f' = \frac{n'}{(n' - n)c} \qquad (3.39)$$

The nodal points (see Section 1.3) must coincide at the centre of curvature of the surface, since a ray through the centre of curvature meets the surface normally and is undeviated; this result also follows from eqn (1.14).

For a mirror with light incident from the left we take $n = -1$ and $n' = 1$ and we find that the two focal lengths are both equal to $1/2c$ and the power is $2c$, where c is again the curvature. As before the principal points are at the mirror surface and the nodal points are at the centre of curvature.

3.8   Gaussian properties of two systems

Our next step is to consider the combination of two general systems of known properties and with a given axial separation. It is convenient to do this by determining the paths of certain paraxial rays through the combined system and obtaining the Gaussian parameters from these paths. Consider first any

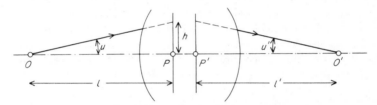

Fig. 3.13. Conjugates and convergence angles

system of power K; its conjugate distance equation referred to the principal planes is (see eqn (3.15))

$$\frac{n'}{l'} - \frac{n}{l} = K \qquad (3.15)$$

and if we multiply through by h, the incidence height of a ray at the principal planes, we obtain an equation in terms of the convergence angles defined in Section 3.3:

$$n'u' - nu = -hK. \qquad (3.40)$$

This is illustrated in Fig. 3.13. If we now let l tend to infinity the image point O' becomes the image side principal focus F' and l' becomes the image side

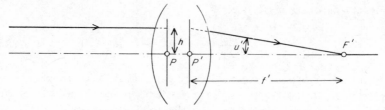

FIG. 3.14. Calculating f' from the path of a paraxial ray

focal length f', according to the definitions of Section 1.3. We have the situation shown in Fig. 3.14 and from eqn (3.40) we have

$$n'u' = -hK \tag{3.41}$$

or,

$$K = -n'u'/h \tag{3.42}$$

and

$$f' = -h/u'. \tag{3.43}$$

The first focal length is easily obtained from the fundamental relationship between the focal lengths, eqn (3.14).

Returning now to the problem of combining two systems in series, let their powers be K_1 and K_2 and let the separation between adjacent principal planes P_1' and P_2 be d, as in Fig. 3.15; this quantity is taken positive if $P_1'P_2$ is a positive distance, i.e. left-to-right; the refractive indices in the initial, intermediate and final spaces are n, n_1' and n'. We now take a ray from the axial object point at infinity at distance h_1 from the axis. By applying eqn (3.40) we have, since $u \equiv 0$ in this case,

$$n_1'u_1' = -h_1K_1. \tag{3.44}$$

By geometry in the intermediate space we have

$$h_2 = h_1 + du_1' \tag{3.45}$$

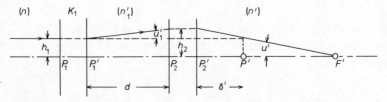

FIG. 3.15. Combining two systems in series; separation referred to the principal points

and by a similar application of eqn (3.40) to the second system,

$$n'u' - n_1'u_1' = -h_2 K_2. \tag{3.46}$$

We can now eliminate h_2 and u_1' between eqns (3.44), (3.45), (3.46), and we obtain

$$n'u' = -h_1\left\{K_1 + K_2 - \frac{d}{n_1'} K_1 K_2\right\}; \tag{3.47}$$

if we compare this with eqn (3.41) we see that the quantity in the brackets on the right-hand side of eqn (3.47) must be K, the power of the combined system, so we have

$$K = K_1 + K_2 - \frac{d}{n_1'} K_1 K_2. \tag{3.48}$$

In order to specify the Gaussian properties completely we have to know the positions of the principal planes P and P′ as well as the power; let P$_2'$P′ be denoted by δ', positive in this sense, and let P$_1$P be δ. From the geometry of the diagram we have

$$\delta' = -(h_2 - h_1)/u', \tag{3.49}$$

$$= -du_1'/u' \tag{3.50}$$

from eqn (3.45); then substituting for u_1' from eqn (3.44) and for u' from eqn (3.42) we have

$$\delta' = -\frac{n'}{n_1'} \cdot \frac{dK_1}{K}. \tag{3.51}$$

If we now reverse the whole procedure, i.e. take a ray which passes through the object side principal focus, we obtain the result

$$\delta = \frac{n}{n_1'} \cdot \frac{dK_2}{K} \tag{3.52}$$

These results can now be applied to some cases of practical importance. For the *thick lens in air*, as in Fig. 3.16, the powers of the individual surfaces are given by the expression on the right of eqn (3.38) and substituting in eqn (3.48) we have for the power

$$K = (n - n_a)\left\{c_1 - c_2 + \left(\frac{n - n_a}{n}\right)dc_1c_2\right\}, \tag{3.53}$$

where n_a is the refractive index of air. This value is at ordinary temperatures and pressures approximately 1·0003 but the universal practice in optical

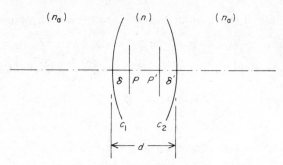

FIG. 3.16. A thick lens in air

design is to specify the indices of glasses as relative to air, not vacuum, so with this convention we have for the thick lens,

$$K = (n - 1)\left\{c_1 - c_2 + \frac{n - 1}{n} dc_1 c_2\right\}, \tag{3.54}$$

where now the relative index is understood. Thus from Section 3.2 we see that the reciprocal of K in eqn (3.54), which is the equivalent focal length, is also the image side focal length. As noted in Section 3.7, the principal planes of a single surface coincide at the surface, so that the positions of the principal planes of the complete lens are given by values of δ and δ' measured as indicated in the figure; here δ' happens to be negative. The explicit formulae are, from eqns (3.51) and (3.52),

$$\delta = -\frac{(n - 1)}{n} \cdot \frac{dc_2}{K}; \qquad \delta' = -\frac{(n - 1)}{n} \cdot \frac{dc_1}{K}. \tag{3.55}$$

If we allow the thickness d to tend to zero we obtain that very useful and important abstraction, the *thin lens in air*; its power is given by

$$K = (n - 1)(c_1 - c_2) \tag{3.56}$$

and its principal planes are at the lens. The concept of the thin lens finds many applications in optical designing, in spite of the manifest impossibility of realizing such a system; the thin lens approximation forms the first stage in the design of many optical systems and the study of the aberrational properties of thin lenses yields useful generalizations about what kinds of optical layout are likely to lead to good designs.

By a further application of eqn (3.48) we obtain the power of two separated thin lenses in air (Fig. 3.17) as

$$K = K_1 + K_2 - dK_1 K_2 \tag{3.57}$$

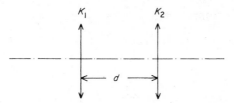

FIG. 3.17. Two separated thin lenses; the symbolic representation for a thin lens as shown is frequently used in texts on optics

and the positions of the principal planes follow from obvious modifications of eqns (3.51) and (3.52).

For two thin lenses in contact it can be seen from eqn (3.57) that the resultant power is the sum of the individual powers; this is one reason for the use of power rather than focal length as a measure of strength in, for example, ophthalmic optics.

It is sometimes useful to re-write some of the results of this section in terms of the separation g between the adjacent foci of the two systems, instead of d, the separation between adjacent principal planes. Paying due regard to signs we have, from Fig. 3.18,

$$d = g + \frac{n_1'}{K_1} + \frac{n_1'}{K_2} \qquad (3.58)$$

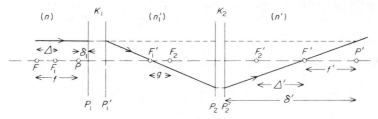

FIG. 3.18. Combining two systems in series; separation referred to the foci

and substituting in eqn (3.48) we obtain for the power of the resultant system

$$K = \frac{-g}{n_1'} \cdot K_1 K_2. \qquad (3.59)$$

In order to obtain compact expressions we find the positions of the principal foci of the resultant system relative to the adjacent foci of the components; referring again to Fig. 3.18 we put $\Delta = F_1F$ and $\Delta' = F_2'F'$ and we consider a ray from the axial object point at infinity; as indicated, F_1' and F' are conjugates on this ray and we have from Newton's conjugate distance relation (eqn (1.8)), with appropriate signs,

$$-g\Delta' = f_2 f_2' \tag{3.60}$$

or

$$\Delta' = \frac{n_1' n'}{g\,K_2^2} = -\frac{n'K_1}{K_2 K}. \tag{3.61}$$

Similarly we obtain

$$\Delta = -\frac{n_1' n}{gK_1^2} = \frac{nK_2}{K_1 K} \tag{3.62}$$

3.9 Thick lenses and combinations of thin lenses

In this section we obtain several interesting properties of lenses and lens systems from the results of Section 3.8; these results will be by way of examples, rather than part of the systematic development of the subject.

If the same increment is added algebraically to the curvatures of a thin lens its power remains the same, as can be seen from eqn (3.56), but its shape, i.e. its cross-section by a plane through the axis, changes and, as will be seen in later chapters, its aberrations change. Figure 3.19 shows how the shape changes with the same increment of curvature at each surface; the process of changing the shape in this way is called *bending* the lens. Since the aberrations

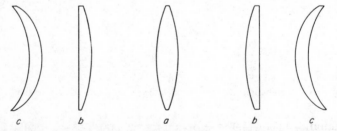

FIG. 3.19. Bending a "thin" lens; the shapes are (a) biconvex, (b) planoconvex, (c) meniscus

change but the Gaussian properties remain constant the amount of bending is a useful variable in balancing aberrations in a complex system. If the lens is not *thin* in the technical sense it can still be bent by adding the same increment to both curvatures but then, as can be seen from eqns (3.54) and (3.55), the power changes and the principal planes shift; however, if the thickness is small compared to the focal length and the radii of curvature, the change in power is small and it is useful to see how the principal planes shift with bending. The distance PP′, sometimes called the *interstitium*, can easily be found from eqn (3.55) to be

$$PP' = d\left\{1 - \frac{1}{n} + \left(\frac{n-1}{n}\right)^2 \frac{dc_1 c_2}{K}\right\}\qquad(3.63)$$

and under the conditions mentioned above it can be seen that this is nearly constant with bending at the value $d(n-1)/n$, i.e. one-third of the thickness of the lens if the conventional approximate value of the refractive index of glass is adopted. However, as can be seen also from eqn (3.55), the principal planes are displaced along the axis roughly proportionally to the increment in curvature. Figure 3.20 shows how this occurs with positive and negative lenses.

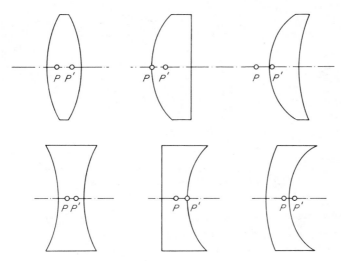

FIG. 3.20. Dependence of the principal points on shape

For some purposes it is desirable to keep the Gaussian effect of a lens constant as it is bent; it would then be necessary to displace the physical lens along the axis in such a way as to keep the principal planes in the same positions relative to the adjacent components. It might also be desirable to add slightly different increments of curvature to the two surfaces so as to keep the power precisely constant and even to allow for the slight change in the interstitium; if this were done then the Gaussian properties of the rest of the system would not be changed and, as will be seen in Chapter 8, the primary aberrations of the rest of the system would also remain constant. However, it is not usual to go to such refinements in bending a lens.

The value of the refractive index does not greatly affect the above discussion for lenses of moderate thickness, i.e. thickness small compared to either radius of curvature and to the equivalent focal length. Very thick lenses are occasionally of interest; for example a sphere of radius r has power $2(n - 1)/nr$ and its principal points coincide at the centre.† A concentric meniscus as in Fig. 3.21 is also "very thick" in this sense; its power is $-(n - 1)dc_1c_2/n$ and its principal points are at the centre of curvature. If the curvature of the second surface is slightly decreased so that the radii are related by

$$r_2 = r_1 - \frac{n - 1}{n} d \qquad (3.64)$$

the power decreases to zero and the meniscus is afocal.

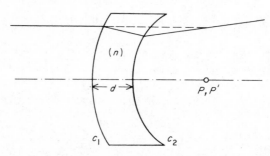

FIG. 3.21. A concentric meniscus lens; the principal points coincide at the common centre of curvature of the surfaces

Two lenses with adjacent foci coincident form an afocal system, as in Fig. 3.7, with angular magnification $-K_2/K_1$, as can be seen from eqns (3.59) and (3.31). A system of three thin lenses arranged as in Fig. 3.22 has unit angular, transverse and longitudinal magnifications, by the formulae of Section 3.4; thus it merely transports a real image a fixed distance along the axis, as indicated by the ray in the figure. A sequence of such systems is used for transporting laser beams without beam spreading and this is also the principle of image transport in periscopes. The electron-optical lens system formed by a uniform axial magnetic field (long solenoid) has essentially the same Gaussian properties, apart from image rotation.

† In fact the principal points of any system with spherical symmetry and with object and image spaces of the same refractive index coincide at the centre of symmetry; this can be seen from the fact that the principal points in this case coincide with the nodal points and the nodal points must be at the centre because a ray aimed at the centre is undeviated by the system.

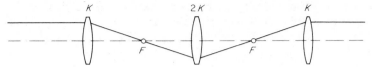

FIG. 3.22. An afocal system of magnification unity

The power and the equivalent focal length of an optical system are the basic measures of its strength, ability to magnify, form bright images, etc. However, for practical purposes it is also useful to define other quantities; the *back focal length* is defined as the distance from the last surface of the system to F′ and the *back vertex power* is the reciprocal of this quantity. There are similar definitions relative to F. These definitions are used chiefly in connection with spectacle lenses and photographic objectives and so it is generally assumed that the medium on both sides of the optical system is air. An interesting example in the Gaussian optics of lens systems is provided by the theory of an instrument for measuring the vertex power of a lens, sometimes called a vertometer. Referring to Fig. 3.23, a lens of known power K is set up with a graticule or marker which can be moved axially in its object space; a telescope adjusted for infinity is set up to view the movable graticule through the lens; thus when the graticule appears in focus it must be at the object side

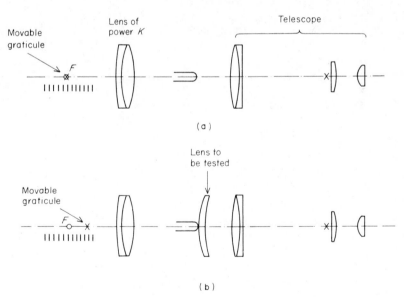

FIG. 3.23. Principle of vertometer (a) set for zero power; (b) measuring the vertex power of a converging lens

principal focus F of the lens. There is a rounded contact at F′ and the lens to be measured is placed with its vertex touching the contact; the graticule must then be displaced through a distance z, say, along the axis in order to bring it into focus, as in the lower diagram. It can be seen that the graticule appears at a distance $z′$ from the vertex of the lens being measured which is the reciprocal of the vertex power, and by Newton's conjugate distance relation eqn (1.8) we have

$$zz′ = -1/K^2, \tag{3.65}$$

thus the vertex power is obtained as $-zK^2$, with due attention to signs.

A final example is provided by the *field lens*, a concept with a variety of applications. One application is to the positioning of the exit pupil in optical systems which are to be used visually, e.g. microscopes; since all the principal rays pass through the centres of the pupils it can be seen that the eye must be placed with *its* entrance pupil at or near the exit pupil of the instrument, as in Fig. 3.24, in order to see the whole field of view.† It is thus necessary to ensure

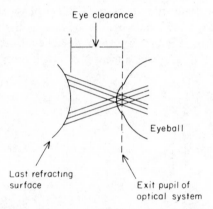

FIG. 3.24. The exit pupil in a visual instrument

that the exit pupil is beyond the last surface of the system and that there is adequate eye-clearance, usually between 10 mm and 20 mm. A field lens is used to bring the exit pupil to a suitable position and also to produce a larger angular field in such cases. Consider, for example, the refracting telescope as in Fig. 3.7; we can re-draw this to show an off-axis or oblique pencil, as in Fig. 3.25; usually the front lens or *objective* is itself the entrance pupil and aperture stop, so that the exit pupil is formed near the second principal focus of the eyepiece; a low-power eyepiece may thus produce a very remote pupil

† The same applies, of course, if two instruments are to be used in series.

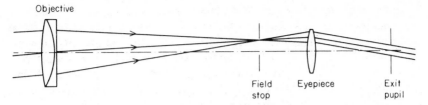

FIG. 3.25. Refracting telescope with single lens eyepiece; the exit pupil is at a large distance from the eyepiece

and so, because of the finite diameter of the eyepiece, a restricted angular field of view. Now suppose a thin lens of positive power, the field lens, is placed at the intermediate real image, where there would normally be a field stop, as in Fig. 3.26; this has no effect on the object–image conjugates and magnification

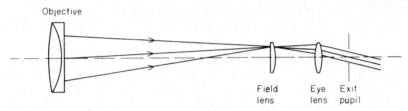

FIG. 3.26. The effect of a field lens on the position of the exit pupil

but the effect is to deflect the oblique pencil towards the axis, thus decreasing the eye-clearance and also decreasing the used diameter of the last component. Most eyepieces except certain very high power systems can be seen to have two sections which may be called the field lens and the eye-lens; in fact each may contain more than one component but the functions of pupil and object imagery are roughly distributed as indicated here.

3.10 Paraxial raytracing

The Gaussian properties of systems containing more than two surfaces or thin lenses can in principle be obtained by successive applications of the formulae in Section 3.8 for combining two systems but in practice the results are clumsy and expressions of this kind are not used in designing and analysing optical systems. Suppose that it is required to find the power of a system of given properties, i.e. the curvatures, refractive indices and thicknesses are all given; the power could be found if we knew the final convergence angle of a paraxial ray of given incidence height from the axial object point at infinity, using eqn (3.42), as was indicated in Fig. 3.14. In that case the result was obtained as an algebraical expression, eqn (3.48), involving the two component systems or surfaces but in the present case there may be any number of surfaces and so we track the path of the ray surface by surface, using *numbers* instead of symbols. This process of finding a ray path in terms of the numerical values of the incidence heights and convergence angles at each surface in turn is called *raytracing* and it is of fundamental importance in optical design; this is because the results obtained are used in aberration calculations in addition to yielding the Gaussian properties.

We have first to number the surfaces and spaces consistently. Suppose there are k refracting or reflecting surfaces; we number them from the left, using the number as a subscript, so that c_j is the curvature of the jth surface; the refractive indices of the media preceding and following c_j are n_j and n'_j respectively and the axial distance from c_j to c_{j+1} is d'_j, taken as positive if c_{j+1} is to the right of c_j, as is usual; the symbol d_{j-1} would denote the distance from c_j to c_{j-1} and it would usually be negative, but this is not often required. There is some redundancy of symbols, since $n'_j = n_{j+1}$ and $d'_j = -d_{j+1}$.† Figure 3.27 shows an optical system with ray segments numbered according to this scheme.

It can be seen from Fig. 3.27 that a ray segment incident at surface j is specified by its incidence height h_j and its convergence angle u_j; the refracted ray is specified by h_j and u'_j so that the incidence heights and convergence angles are in a sense generalized co-ordinates of the ray as it passes through the optical system. We then have from eqn (3.40)

$$n'_j u'_j - n_j u_j = -h_j K_j, \tag{3.66}$$

where

$$K_j = (n'_j - n_j)c_j; \tag{3.67}$$

also by analogy with eqn (3.45)

$$h_{j+1} = h_j + d'_j u'_j. \tag{3.68}$$

The powers, refractive indices, etc., are, of course, given; thus if we start from u_j and h_j we can apply eqns (3.66) and (3.68) in succession to obtain u_{j+1} and h_{j+1}, and so on through the system; eqn (3.66) is called the refraction equation and eqn (3.68) is the transfer equation, both terms having obvious derivations. These two are thus a set of recurrence or recursion formulae, to be applied *numerically*. For example, if we begin with a ray from the axial object point at infinity, for which h_1 is chosen at will and $u_1 = 0$, we obtain h_k and u'_k at the final surface and we then, according to eqn (3.42), obtain the

---

† It is also possible to number spaces rather than surfaces, in which case the surface between spaces j and $j + 1$ would have some such symbol as $c_{j,\,j+1}$ or perhaps simply c_{j+1} as denoting rather arbitrarily the surface following space j. It is essential to have a consistent notation but the choice is a matter for the individual and no special claim is made for that described here.

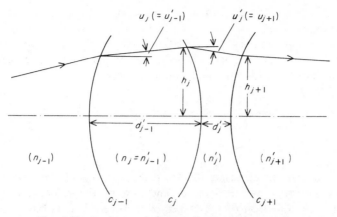

FIG. 3.27. A scheme of notation for paraxial raytracing

power:

$$K = \frac{n'_k u'_k}{h_1}. \tag{3.69}$$

Similarly the equivalent focal length and the back focal length are given by

$$\text{efl} = -h_1/n'_k u'_k, \qquad \text{bfl} = -h_k/u'_k. \tag{3.70}$$

Other applications of the raytracing process will be described later. We now consider how the choice should be made of the initial value of h_1 to be used in a raytrace to find the power. Since this whole process is based on results which apply only to the Gaussian region it might at first be supposed that h_1 should be chosen small enough to restrict all the incidence heights to the Gaussian region. In fact we have not yet estimated the non-Gaussian terms (i.e. the primary aberrations, see Chapters 7 and 8) and we therefore cannot do this accurately, but suppose for the moment that a suitably small value of h_1 has been chosen. The raytracing equations (eqns (3.66) and (3.68)) are linear in the variables h and u and also the end results required, powers or focal lengths, are of zero dimension in h and u jointly; thus if we multiply h_1 by a large numerical factor, then all the succeeding h_j and u_j will merely be scaled by the same factor and the powers and focal lengths will be unchanged. Thus it can be seen that paraxial raytracing can be carried out with any arbitrarily chosen incidence height of the incoming ray and the results are still valid for the above purposes. The incidence height which is generally chosen is that corresponding to the radius of the entrance pupil; when this is done the primary aberrations, calculated as in Chapter 8, also have values appropriate to the edge of the pupil.

The question of the choice of incidence heights for a paraxial raytrace can also be considered in a different way. It may be said that what we are really doing in tracing a ray from, say, the axial object at infinity to the second principal focus is applying in succession the conjugate distance equation for a single surface (eqn (3.38)) and a simple transfer equation,

$$\left.\begin{array}{r} \dfrac{n'_j}{l'_j} - \dfrac{n_j}{l_j} = K_j \\[2mm] l_{j+1} = l'_j - d'_j \end{array}\right\} \tag{3.71}$$

and that the raytracing equations are really versions of these obtained by multiplying through by the factors h_j and u'_j respectively. This point of view gets round the question of the choice of a value of h_1 within the paraxial region but on the other hand, eqns (3.71) only give the back focal length directly and it is necessary to introduce incidence heights somewhere in order to obtain the equivalent focal length or the power.

The raytracing equations may be written down at length as follows:

$$\left.\begin{array}{r} n'_1 u'_1 - n_1 u_1 = -h_1 K_1 \\ h_2 = h_1 + d'_1 u'_1 \\ n'_2 u'_2 - n_2 u_2 = -h_2 K_2 \\ h_3 = h_2 + d'_2 u'_2 \\ \cdots \cdots \cdots \cdots \\ n'_k u'_k - n_k u_k = -h_k K_k \end{array}\right\} \tag{3.72}$$

As stated at the beginning of this section, it is not usual to carry out an algebraic elimination of the intermediate variables to obtain u'_k as a function of h_1 and there is no situation which requires this in present-day optical design techniques. This elimination has been done by means of continued fractions and it is also possible to express eqns (3.72) in matrix form and use a continued matrix product, but this does not seem very useful. An equation which is sometimes useful is obtained by simply adding together all the refraction equations, giving

$$n'_k u'_k - n_1 u_1 = -\{h_1 K_1 + h_2 K_2 + \cdots + h_k K_k\}; \tag{3.73}$$

this equation applies equally to thin lenses in air with powers $K_1, K_2, \ldots$ provided the refractive indices are set equal to unity, and in this form it is used in situations where the aberrations of thin lens systems are to be minimized.

A raytrace according to eqns (3.72) from one axial object point specified by h_1 and u_1 to the conjugate in image space yields complete information about the Gaussian image formation at these conjugates, since the transverse magnification is immediately given according to eqn (3.21) by

$$\frac{\eta_k'}{\eta_1} = \frac{n_1 u_1}{n_k' u_k'}. \tag{3.74}$$

If the object is at infinity, the form of the Lagrange invariant for star space gives a relation between the angular subtense β_1 of the object and the size η_k' of the image,

$$n_k' u_k' \eta_k' = -n_1 h_1 \beta_1 \tag{3.75}$$

in terms of the raytrace results.

There is no need to trace further rays between these conjugates since according to Section 3.1 there is ideal image formation in the Gaussian region.

Many optical systems are designed for use with the object at infinity, e.g. almost all photographic objectives. As a final example we note that eqn (3.75), together with eqn (3.42) of Section 3.8, illustrates a method of measuring the power of such systems. By eliminating u_k'/h_1 from these two equations we have

$$K = \frac{-n_1 \beta_1}{\eta_k'} \tag{3.76}$$

thus if the size of image η_k' corresponding to an object of angular subtense β_1 is found, the power is obtained. This, the "focometer" method, is considered to be one of the most accurate methods of measuring power or efl; it has even been suggested as a way of defining efl for an industrial standard.

4   Finite Raytracing

4.1   Finite rays

So far we have considered rays and image formation only in the paraxial region and the "raytracing" of Section 3.10 is really a process only valid in this region, although as a numerical convenience we gave fictitiously large values to the incidence heights and convergence angles. Having established the Gaussian image formation for a certain conjugate pair in this way it is possible to conceive of a process in which rays would be traced from a chosen point O_1 of the object plane without making the approximations of Gaussian optics, i.e. exactly according to Snell's law (Section 2.2), and their intersections with the Gaussian image plane found. These intersections would in general not coincide at O_1', the Gaussian image of O_1, and the scatter in the ray intersections would be a manifestation of the aberrations of the optical system. Rays traced in this way may be called *finite* rays, to distinguish them from paraxial rays which are in the mathematical sense infinitesimal, and the process of tracing them is known as finite raytracing. This is an essential tool of the optical designer and it is applied not only to axisymmetric systems but to all others including, e.g. systems including diffractive optical elements such as hologram lenses and systems with toric surfaces. Finite raytracing is used at the stage when the general form of the optical system is known but the aberrations have to be reduced to acceptable tolerance levels.

Finite raytracing is essentially an iterative sequence of two operations; one of these, transfer, is simply taking a ray from where it leaves one optical surface to where it meets the next and it thus involves only the geometry of straight lines and surfaces of appropriate shape. The other operation, usually called refraction, is to find the direction of the ray after it has passed through or been reflected or diffracted by the surface; the calculation involves either Snell's law or a generalized form of it appropriate to diffractive optical elements. In this chapter we shall deal only with refracting and reflecting surfaces of axial symmetry; toric surfaces and diffractive components will be dealt with in Chapter 5. The actual amount of numerical computation can, of course, be quite considerable and this led in the past to the development of many systems of formulae, each aimed at reducing the numerical work according to the particular computing aids available, i.e. logarithmic and trigonometric tables and various forms of desk calculating machine. It was also the case that most work was done with rays lying in a plane through the optical axis (meridian rays); rays skew to the axis were not often traced since much more work was involved than in tracing meridian rays. The universal availability of electronic computers has changed this situation completely since a skew raytrace can now be carried through a complete system in a small fraction of a second. At the time of writing the fastest machines will trace skew finite rays at the rate of about 15000 ray-surfaces per second, whereas a skilled human took perhaps 20 min for one ray-surface using log tables and perhaps 5 min with a desk calculator. We therefore describe first the universally used system for tracing skew rays through spherical surfaces and we then discuss modifications suitable for non-spherical surfaces; finally we shall give a method suitable for tracing meridian rays through spherical surfaces by trigonometrical tables. In all cases there are the two operations of refraction and transfer, applied alternately, just as described in Section 3.10 for paraxial raytracing. We deduce and discuss the formulae in detail in the rest of this chapter; the equations are collected for reference in Appendix A.

4.2 Snell's law for skew rays

It will be recalled from Chapter 2 that Snell's law can be written in the vector form

$$n'(r' \wedge n) = n(r \wedge n), \tag{2.3}$$

where n is a unit vector along the normal at the point of incidence and r and r' are unit vectors along the incident and refracted rays. If this equation is multiplied vectorially by n we obtain

$$n'(r' - n(r' \cdot n)) = n(r - n(r \cdot n)). \tag{4.1}$$

This can be expanded in scalar form by setting (L, M, N), (L', M', N') and (α, β, γ) as the components of r, r' and n respectively, so that these quantities are direction cosines, giving

$$\left. \begin{array}{c} n'L' - nL = k\alpha \\ n'M' - nM = k\beta \\ n'N' - nN = k\gamma \end{array} \right\}, \tag{4.2}$$

where

$$\begin{aligned} k &= n'(r' \cdot n) - n(r \cdot n) \\ &= n' \cos I' - n \cos I \end{aligned} \tag{4.3}$$

and I and I' are the angles of incidence and refraction.

Equations (4.2) are in a suitable form for use in the systems for tracing skew rays which we shall develop.

4.3 Transfer between spherical surfaces

Let a finite ray leave a surface (the "previous" surface) with direction cosines (L, M, N) and let its point of incidence P_{-1} at this surface have co-ordinates (x_{-1}, y_{-1}, z_{-1}) in the local coordinate system, i.e. the system with its origin at the vertex of this surface. The transfer problem is to determine P, the point of incidence of the ray on the current surface (Fig. 4.1), it being assumed that (L, M, N) and the coordinates of P_{-1} are known from previous steps. Let c be the curvature of the current surface and let d be the axial

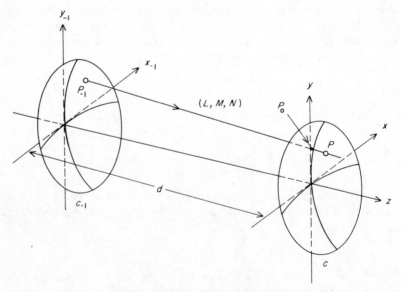

Fig. 4.1. Coordinates and notation for finite ray transfer

separation. It is convenient to take the calculation in two steps, first finding the coordinates of the point P_0 in which the ray meets the x–y plane for the current surface. If the coordinates of P_0 are $(x_0, y_0, 0)$ we have

$$\left. \begin{array}{l} x_0 = x_{-1} + \dfrac{L}{N}(d - z_{-1}) \\[3mm] y_0 = y_{-1} + \dfrac{M}{N}(d - z_{-1}) \end{array} \right\}, \tag{4.4}$$

by the standard properties of direction cosines.

Now let the coordinates of P be (x, y, z) and let the length of the segment P_0P be Δ. We have

$$\left. \begin{array}{l} x = x_0 + L\Delta \\ y = y_0 + M\Delta \\ z = N\Delta \end{array} \right\}. \tag{4.5}$$

The further condition, that P lies on a surface of curvature c which is tangent to the x–y plane at the origin, can be put in the form

$$z = \tfrac{1}{2}c(x^2 + y^2 + z^2). \tag{4.6}$$

We can eliminate x, y and z between eqns (4.5) and (4.6) to obtain a quadratic in Δ:

$$\tfrac{1}{2}c\{(x_0 + L\Delta)^2 + (y_0 + M\Delta)^2 + N^2\Delta^2\} = N\Delta, \tag{4.7}$$

or

$$c^2\Delta^2 - 2c\{N - c(Lx_0 + My_0)\}\Delta + c^2(x_0^2 + y_0^2) = 0. \tag{4.8}$$

Let

$$F = c(x_0^2 + y_0^2) \tag{4.9}$$

and

$$G = N - c(Lx_0 + My_0), \tag{4.10}$$

then the solution of the quadratic is

$$c\Delta = G \pm \sqrt{G^2 - cF} \tag{4.11}$$

By considering the special case when x_0 and y_0 tend to zero, i.e. the ray is incident at the vertex of the refracting surface, we see that the negative sign is to be taken in eqn (4.11). Also it is generally the case that $|cF|$ is much less than G^2, since the latter is of order unity whereas cF is of the order of the square of the angle subtended at the centre of curvature of the refracting surface by the arc joining the point of incidence to the vertex, and this angle is usually less than 0·2 rad; in these circumstances a direct numerical evaluation of eqn (4.11) would give $c\Delta$ as the difference of two nearly equal numbers and a loss of numerical precision would occur. This is avoided by multiplying and dividing the right-hand side of eqn (4.11) (with negative sign) by the quantity $G + \sqrt{G^2 - cF}$. The result is

$$\Delta = \frac{F}{G + \sqrt{G^2 - cF}}, \tag{4.12}$$

a form which involves no loss of significant figures when used numerically.

The value of Δ obtained from eqn (4.12) can be substituted in eqn (4.5) to give (x, y, z), the coordinates of the point of incidence; this completes the transfer process.

4.4   Refraction through a spherical surface

To obtain refraction formulae we need to know the components of the unit normal at the point of incidence, in order to apply eqn (4.2.). It can easily be seen that for a spherical surface these are given by

$$(\alpha, \beta, \gamma) = (-cx, -cy, 1 - cz), \tag{4.13}$$

where (x, y, z) is the point of incidence and c is the curvature. Substituting in eqn (4.2) we have as the refraction equations

$$\left.\begin{array}{l} n'L' = nL - Kx \\ n'M' = nM - Ky \\ n'N' = nN - Kz + n' \cos I' - n \cos I \end{array}\right\}, \tag{4.14}$$

where

$$K = c(n' \cos I' - n \cos I). \tag{4.15}$$

We have, of course, to find $\cos I$ and $\cos I'$ before these can be applied. By scalar multiplication we obtain

$$\begin{aligned} \cos I &= \mathbf{r} \cdot \mathbf{n} \\ &= N - c(Lx + My + Nz) \\ &= N - c\{L(x_0 + L\Delta) + M(y_0 + M\Delta) + N^2\Delta\} \\ &= N - c\Delta - c(Lx_0 + My_0), \end{aligned}$$

or

$$\cos I = +\{G^2 - cF\}^{\frac{1}{2}}, \tag{4.16}$$

by eqns (4.10) and (4.11).

Thus cos I has already been found in the transfer process.

We find cos I' by application of Snell's law in the form of eqn (2.2):

$$n' \cos I' = +\{n'^2 - n^2(1 - \cos^2 I)\}^{\frac{1}{2}}. \qquad (4.17)$$

This completes the refraction process. The raytrace then proceeds as an iterative sequence of transfers and refractions, using the equations of this and the preceding sections alternately.

Numerical checks are sometimes useful, particularly when running a new raytracing program. A check on the transfer process is provided by the fact that the coordinates of the point of incidence must satisfy the equation of the refracting surface, which is eqn (4.6); the refraction process can be checked by the requirement that the sum of the squares of the direction cosines must equal unity:

$$L'^2 + M'^2 + N'^2 = 1. \qquad (4.18)$$

It is worth noting that the finite raytracing equations for transfer and refraction (eqns (4.4) and (4.14)) reduce directly to the corresponding paraxial forms (eqns (3.72)) when the appropriate paraxializing approximations are made; these are

$$N \rightarrow 1, \quad z \rightarrow 0, \quad \Delta \rightarrow 0, \quad \cos I \rightarrow 1, \quad \cos I' \rightarrow 1. \qquad (4.19)$$

The quantity K of eqn (4.15) then reduces to the Gaussian power, $(n' - n)c$ for which we have intentionally used the same symbol; it is therefore called the generalized power and it is a function of the angle of incidence at the surface.

4.5 Beginning and ending a raytrace

The procedures for starting a raytrace and finishing it depend very much on the nature of the information required from the trace; we list first a few possibilities in increasing order of complexity.

(a) We may wish to trace simply a principal ray at a given field angle, in order to determine distortion by comparing the point at which it meets the image plane with the Gaussian prediction.

(b) We may wish to trace a group of rays from an axial object point in order to evaluate the aberration of the axial pencil (spherical aberration).

(c) A similar pencil from an off-axis object point may be required, in which case skew rays have to be included.

(d) In both (b) and (c) above it may be required to trace enough rays to enable a polynomial to be constructed to represent the ray aberrations; this polynomial might be needed to facilitate some kind of deeper evaluation of the system, e.g. a computation of the optical transfer function (see Chapter 13).

These requirements and similar ones occurring on different occasions all involve tracing a pencil of rays from a given object point, which will frequently be at infinity, through the system in such directions that they meet the aperture stop or one of the pupils in a regular grid or mesh. Usually it is convenient to put the regular grid at the entrance pupil and then the pattern of ray intersections at the exit pupil will be distorted by pupil aberrations, but this usually does not matter. Thus if the object point is at infinity the raytrace starts with a set of parallel rays, direction cosines $(0, M_0, N_0)$ leaving the pupil, as in Fig. 4.2, with a grid spacing depending on the type of system, the kinds of aberrations expected and the nature of the subsequent calculations.

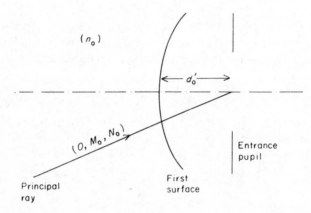

FIG. 4.2. Starting a raytrace from an object point at infinity in the direction $(0, M_0, N_0)$

It is convenient to treat the pupil as a plane *ghost surface*, i.e. a surface with no refractive index change; the first step is then a transfer operation from the pupil on to the first refracting surface at a distance d_0'; usually the entrance pupil is physically inside the system, so that d_0' is negative; but optically it is, of course, in object space and it is on the object side of the first refracting surface.

The raytrace is completed by a final transfer from the last refracting surface to the Gaussian image plane, or other chosen image surface, and the coordinates of the ray intersections are compared with the Gaussian image position to give the transverse ray aberrations; this is discussed in more detail in Chapter 8. If the object is at a finite distance it is possible to start at the object plane with a pencil of rays from a single object point having suitably chosen direction cosines and to transfer direct to the first refracting surface. However, if the object plane is at a considerable distance, this process leads to loss of numerical accuracy and it is better to start at the entrance pupil with a grid of rays having varying direction cosines. There is inevitably a loss of significant figures in transferring over a long distance by eqn (4.4) and it is therefore usual to arrange that a raytrace starts at the longer conjugate, when the transfer is avoided as above, even if this means reversing the system. Thus for example projection lenses are designed by tracing rays from the projection screen (usually assumed to be at infinity) to the plane of the object transparency and microscope objectives are traced from the long conjugate to the short. This reversal actually facilitates the interpretation of the aberration patterns.

When oblique pencils (i.e. pencils from an off-axis object point) are traced it is usually necessary to allow for vignetting. In a raytracing program this is done by checking at each surface and at the aperture stop whether the ray falls inside the predetermined diameter of the lens or stop; for some purposes it is necessary to determine, by interpolation, rays which just graze the vignetted pupil. Thus the entrance and exit pupil grids might appear as in Fig. 4.3, where the effect of pupil aberrations is seen as a distortion of the ray grid. Even more complex effects can occur in reflecting systems; the axial

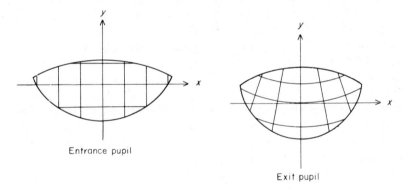

FIG. 4.3. Vignetted and distorted pupils

pupil normally has a central obstruction in such systems and the oblique pupil can appear as in Fig. 4.4. Clearly the ray grid must be chosen with some care in such cases if it is to represent a point image reasonably well.

4.6 Non-spherical surfaces

Most optical surfaces are spherical (or plane) because a spherical surface is the easiest to produce; it is formed when two surfaces wear together to a

Axial pupil   Oblique pupil

F_{IG}. 4.4. Vignetting in a reflecting system; the hatched areas are those obstructed by
reflections of mirrors

uniform contact and it is therefore obtained naturally by a uniform grinding
or polishing action.† However, non-spherical surfaces are necessary for some
purposes and in this section we show how rays are traced through them. There
is an immediate difficulty in specifying an aspheric: a sphere has only one
parameter, its curvature, but an aspheric of revolution can have an infinite
number of parameters, these being, say, the coefficients in the power series
specifying its meridian section. Alternatively we can choose a surface of a
definite shape, say a paraboloid of revolution; certain such surfaces, namely
the quadrics of revolution, appear as the solutions of particular optical
problems and in some cases methods have been developed for direct generation, by linkages, of the surfaces. There is a further attraction to the special
case of quadrics, that the transfer and refraction operations are in terms of
simple closed formulae which are very similar in form to those obtained in
Sections 4.3 and 4.4 for spherical surfaces. On the other hand some form of
iterative calculation is necessary for transfer in the case of an aspheric expressed as a power series since this involves the solution of an equation of
degree equal to the power of the highest term in the series. The possibilities of
aspheric surfaces in optical design have not so far been fully explored, but
certainly surfaces expressed as quadrics and as general power series up to
about the tenth degree have both been found useful, so we shall deal with these
two cases.

It is necessary to have convenient expressions for the direction cosines of
the normal to a surface so that the general refraction equations can be used
(eqn (4.2)), and we first give these expressions. Let the equation of the surface,
which need not for this purpose be of revolution symmetry, be

$$F(x, y, z) = 0; \tag{4.20}$$

† Unfortunately this is not strictly true; departures from true sphericity which are not
negligible on an optical scale usually occur and these have to be corrected by careful control
of the polishing process. These variations are not reproducible and cannot therefore be used
to generate chosen non-spherical shapes.

for a neighbouring point $(x + \delta x, y + \delta y, z + \delta z)$ which is also on the surface we have

$$F(x + \delta x, y + \delta y, z + \delta z) = 0 \qquad (4.21)$$

or, by Taylor's theorem,

$$F(x, y, z) + \frac{\partial E}{\partial x} \delta x + \frac{\partial F}{\partial y} \delta y + \frac{\partial F}{\partial z} \delta z + \cdots = 0 \qquad (4.22)$$

Thus from eqn (4.22) we see that the vector

$$\left(\frac{\partial F}{\partial x}, \frac{\partial F}{\partial y}, \frac{\partial F}{\partial z} \right) \qquad (4.23)$$

is perpendicular to $(\delta x, \delta y, \delta z)$ and since the latter is restricted to lie in the surface the expression (4.23) must be a vector along the normal. Therefore the direction cosines of the normal are

$$(\alpha, \beta, \gamma) = \frac{-\left(\dfrac{\partial F}{\partial x}, \dfrac{\partial F}{\partial y}, \dfrac{\partial F}{\partial z} \right)}{\left\{ \left(\dfrac{\partial F}{\partial x} \right)^2 + \left(\dfrac{\partial F}{\partial y} \right)^2 + \left(\dfrac{\partial F}{\partial z} \right)^2 \right\}^{\frac{1}{2}}}. \qquad (4.24)$$

This result will be used in the derivation of refraction equations in the following sections.

4.7 Raytracing through quadrics of revolution

For raytracing purposes it is convenient to represent the general quadric of revolution by the equation

$$z = \tfrac{1}{2}c(x^2 + y^2 + \varepsilon z^2). \qquad (4.25)$$

It is easily seen that this is a surface of revolution about the z-axis, passing through the origin and having curvature c at that point. The parameter ε determines the asphericity as follows:

$$
\begin{aligned}
\varepsilon &> 1, \quad \text{prolate ellipsoid,} \\
\varepsilon &= 1, \quad \text{sphere,} \\
0 < \varepsilon &< 1, \quad \text{oblate ellipsoid,} \\
\varepsilon &= 0, \quad \text{paraboloid,} \\
\varepsilon &< 0, \quad \text{hyperboloid,}
\end{aligned}
$$

as may easily be determined by putting the equation into the appropriate canonical form. In terms of the eccentricity e, which is the more usual parameter, we have

$$
\left.\begin{array}{l}
e = \sqrt{1 - \varepsilon}, \quad \varepsilon > 0 \\
e = \sqrt{1 + \varepsilon}, \quad \varepsilon < 0
\end{array}\right\}. \tag{4.26}
$$

For our purposes it is convenient to have a form such as eqn (4.25) in which one parameter varies continuously to give a range of asphericities while keeping the paraxial curvature constant; the definition of the eccentricity changes according to whether the surface is an ellipsoid or a hyperboloid, so that it is an inconvenient parameter.

Sometimes yet another parameter is used which is equal to $\varepsilon - 1$. This would put eqn (4.25) in the form of a sphere plus an extra aspheric term. Equation (4.25) does not include the cone although cones are degenerate quadrics; for some purposes it would be possible to approximate to the effect of a cone by using a hyperboloid with a very large value of ε.

We can now obtain transfer equations for the quadric. Proceeding exactly as in Section 4.3 we find the point $P_0(x_0, y_0, 0)$ at which the incoming ray meets the x–y plane, i.e. the tangent plane to the surface where it meets the optical axis. Then let $P(x, y, z)$ be the point of incidence of the ray at the surface and let the distance P_0P be Δ; as in Section 4.3 we have

$$
\left.\begin{array}{l}
x = x_0 + L\Delta \\
y = y_0 + M\Delta \\
z = N\Delta
\end{array}\right\} \tag{4.27}
$$

and Δ is found as the solution of the quadratic obtained by eliminating x, y and z between these equations and eqn (4.25). The procedure, including selection of the appropriate root, is as in Section 4.3 and the result is

$$
\Delta = \frac{F}{G + \{G^2 - cF(1 + (\varepsilon - 1)N^2)\}^{\frac{1}{2}}}, \tag{4.28}
$$

where as before

$$
\left.\begin{array}{l}
F = c(x_0^2 + y_0^2) \\
G = N - c(Lx_0 + My_0)
\end{array}\right\}. \tag{4.29}
$$

It can easily be seen that the result for a sphere is recovered if ε is put equal to unity. The value of Δ is substituted in eqns (4.27) to complete the transfer.

For the refraction calculation we again have first to find the cosine of the angle of incidence of the ray; by applying eqns (4.24) to the equation of the quadric we obtain for the direction cosines of the normal

$$
(\alpha, \beta, \gamma) = \frac{(-cx, -cy, 1 - c\varepsilon z)}{\{1 - 2c(\varepsilon - 1)z + c^2\varepsilon(\varepsilon - 1)z^2\}^{\frac{1}{2}}} \tag{4.30}
$$

By scalar multiplication with the direction cosines of the ray we have

$$\cos I = \frac{N - c(Lx + My + N\varepsilon z)}{\{1 - 2c(\varepsilon - 1)z + c^2\varepsilon(\varepsilon - 1)z^2\}^{\frac{1}{2}}} \tag{4.31}$$

If eqns (4.27) are used to put this expression in terms of x_0 and y_0 it is possible to obtain an expression for $\cos I$ which is analogous to eqn (4.16):

$$\cos I = \frac{\{G^2 - cF(1 + (\varepsilon - 1)N^2)\}^{\frac{1}{2}}}{\{1 - 2c(\varepsilon - 1)N\Delta + c^2\varepsilon(\varepsilon - 1)N^2\Delta^2\}^{\frac{1}{2}}} \tag{4.32}$$

However, this lacks the simplicity of eqn (4.16) and it is probably just as easy to use eqn (4.31). The refraction calculation is completed by applying eqn (4.2).

4.8 The general aspheric surface

According to the argument in Section 4.6 we take as our general aspheric a surface of revolution specified by an equation such as

$$z = \tfrac{1}{2}c(x^2 + y^2) + b_4(x^2 + y^2)^2 + b_6(x^2 + y^2)^3 + \cdots. \tag{4.33}$$

The number of terms taken will depend on the particular problem; coefficients up to b_{12} have been used in some recent work and on the other hand it often turns out that only a single aspheric coefficient is needed, i.e. b_4, together with the curvature term; for example it can happen that the aspheric plate of a Schmidt camera of moderate size can be so represented with adequate precision, although it would be impossible to fit it with a quadric.

It will in general be impossible to find closed formulae for the transfer operation for a general aspheric, since this would involve the solution of an algebraic equation of degree higher than the fourth by closed expressions; it is well known that this is not possible. The transfer process must therefore be carried out by some system of iterative or recursive formulae. Many such have been devised and we give an example here. The process is illustrated by Fig. 4.5; let the incoming ray meet the x–y plane at $P_0(x_0, y_0, 0)$; we find the point P_1 on the surface with the same x and y coordinates as P_0 and then find P_1', the intersection of the tangent plane at P_1 with the ray. Next we find P_2, the point on the surface with the same x and y coordinates as P_1' and then find P_2', the intersection of the tangent plane at P_2 with the ray. The process is repeated until the coordinates of P_{n+1} are the same as those of P_n to the

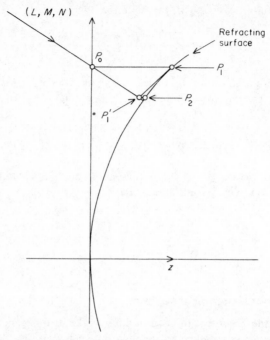

FIG. 4.5. Iterative process for transfer at non-spherical surface

desired number of significant figures. It is, of course, valid for skew rays and it converges rapidly in almost all cases.†

The detailed formulae are as follows. The coordinates of P_0 are obtained as for spherical surfaces by eqn (4.4) and the z coordinate of P_1 is obtained by substituting x_0 and y_0 in the equation of the surface (eqn (4.33)). Then eqns (4.24) are used to obtain the direction cosines $(\alpha_1, \beta_1, \gamma_1)$ of the normal at $P_1(x_0, y_0, z_1)$. The equations of the ray are:

$$\frac{x - x_0}{L} = \frac{y - y_0}{M} = \frac{z}{N} \tag{4.34}$$

and the equation of the tangent plane is

$$\alpha_1(x - x_0) + \beta_1(y - y_0) + \gamma_1(z - z_1) = 0. \tag{4.35}$$

From these we find the coordinates of P_1':

$$z_1' = \frac{N\gamma_1 z_1}{\alpha_1 L + \beta_1 M + \gamma_1 N}, \tag{4.36}$$

† It is possible to construct extreme cases which defeat the process, e.g. the incident ray perpendicular to the z-axis, but such are unlikely to arise in practice.

$$x_2 = \frac{L}{N} z_1' + x_0, \tag{4.37}$$

$$y_2 = \frac{M}{N} z_1' + y_0. \tag{4.38}$$

By substituting x_2 and y_2 in eqn (4.33) we find z_2, the z-coordinate of P_2, and this completes a cycle of the iteration.

For the next cycle we find $(\alpha_2, \beta_2, \gamma_2)$, the normal at P_2, and then the co-ordinates of P_2', the intersection of the tangent plane at P_2 with the ray, are

$$z_2' = \frac{N\{\alpha_2(x_2 - x_0) + \beta_2(y_2 - y_0) + \gamma_2 z_2\}}{\alpha_2 L + \beta_2 M + \gamma_2 N}, \tag{4.39}$$

$$x_3 = \frac{L}{N} z_2' + x_0, \tag{4.40}$$

$$y_3 = \frac{M}{N} z_2' + y_0. \tag{4.41}$$

By substituting x_3 and y_3 in eqn (4.33), we obtain z_3 and the second cycle is completed. The equations are the same for further cycles, except that the subscripts 2 and 3 are changed appropriately.

The process is continued until (x_n, y_n, z_n) repeat to the required number of significant digits; as with all iterative calculations, it has the agreeable property of being self-checking, and a numerical error at any stage only causes some delay in reaching the correct result.

For the refraction calculation we use the final values of the direction cosines of the normal to obtain $\cos I$:

$$\cos I = \alpha L + \beta M + \gamma N \tag{4.42}$$

and it can be seen that this has already been obtained as the denominator of the expression for z_n' (eqn (4.39)). Cos I' follows from eqn (4.17) and then eqn (4.2) is used for the refraction calculation.

It should be emphasized that the above system is only one of many which have been used; the choice is partly dependent on the general approach to design, for example, it may be preferred to express the aspheric surface as a true sphere plus a polynomial representing the aspherizing, or a different kind of iterative process may be more suited to the computer available.

4.9 Meridian rays by a trigonometrical method

As mentioned in Section 4.1, many methods have been used for logarithmic and desk calculator raytracing. For logarithmic calculations probably the most efficient systems were developed by A. E. Conrady in "Applied Optics and Optical Design" (Part I, 1929, Oxford; Part II, 1960, Dover). T. Smith gave several elegant schemes suitable for use with desk calculating machines

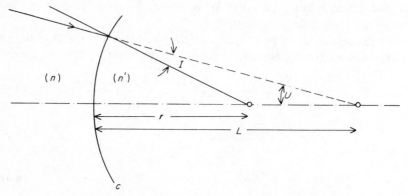

Fig. 4.6. Meridian raytracing by trigonometrical formulae

and more or less abridged trigonometrical tables (*Proc. phys. Soc. Lond.* **27**, 502–9 (1915), **30**, 221–33 (1918), **32**, 252–64 (1920), **33**, 174–78 (1921), **57**, 286–93 (1945)). Many other authors have also given schemes and a complete bibliography would contain dozens of items. These methods are mainly of historical interest now but some of the basic formulae for meridian raytracing are occasionally useful, sometimes in developing closed solutions, e.g. for two-mirror aplanatic systems. We give the simplest system for meridian rays below.

The notation is shown in Fig. 4.6, in agreement with Conrady except that the convergence angle as shown is negative according to our sign convention. The angle of incidence I and the intersection length L are positive as shown. The ray coordinates are L and U. The scheme, which may easily be verified, is as follows:

$$\left. \begin{aligned} \sin I &= -\frac{(L-r)}{r} \sin U \\ n' \sin I' &= n \sin I \\ I' - U' &= I - U \\ L' - r &= -\frac{r \sin I'}{\sin U'} \end{aligned} \right\} \tag{4.43}$$

There are certain situations with this simple scheme where significant figures can be lost, e.g. large L or large r; various modifications involving more computation are used then. Equations (4.43) can be computed with pocket electronic calculators in less than a minute per ray surface, or in a few seconds with programmable calculators.

4.10 Reflecting surfaces

The special case of reflection in finite raytracing is treated as explained in Section 2.2 but for convenience we note here the necessary changes in the principal equations.

Equation (2.3) becomes

$$r' \wedge n = -r \wedge n. \tag{4.44}$$

Equation (4.1) becomes

$$r' + r = 2n(r \cdot n). \tag{4.45}$$

Equations (4.2) become

$$\left.\begin{array}{c} L' + L = 2\alpha \cos I \\ M' + M = 2\beta \cos I \\ N' + N = 2\gamma \cos I \end{array}\right\} \tag{4.46}$$

and eqns (4.14) become

$$\left.\begin{array}{c} L' + L = -kx \\ M' + M = -ky \\ N' + N = -kz + 2 \cos I \end{array}\right\} \tag{4.47}$$

with

$$k = 2c \cos I. \tag{4.48}$$

4.11 Failures and special cases

The general raytrace method for spherical surfaces described in Sections 4.3 and 4.4 is numerically well behaved in the sense that no special cases can occur of large losses of significant figures due to e.g. taking the difference between two nearly equal numbers. However, it is worth noting when failures occur; these are in the transfer operation when the ray misses the next surface completely and in refraction when there is (unintended) total internal reflection at the surface.

The first failure is signalled by a negative value of $G^2 - cF$ in eqn (4.11) or (4.12), meaning that $\cos I$ from eqn (4.16) would be imaginary; this should be appropriately flagged in a raytracing program.

The second case, total internal reflection, is correspondingly signalled by a negative value of the square root in eqn (4.17), indicating that $\sin I'$ exceeds unity. It could happen that the total internal reflection is intended; if this applies to the whole used area of the surface it may simply be treated as a reflecting surface in a medium of appropriate refractive index; in the unlikely event that part of the surface is to transmit and part to reflect it is probably best to determine the boundary and write a special subroutine for the surface. (Certain prisms are used in this way but they can usually be dealt with by developing them into plane parallel plates, as indicated in the introduction to Chapter 1.)

5 Finite Raytracing through Non-symmetrical Systems

Once we depart from axial symmetry the possibilities are endless. Alternatively we might say that finite raytracing through any kind of surface is covered in principle by Chapter 4, provided that appropriate modifications to the algorithms for transfer and for finding the normal to the next surface are made; this, however, is not strictly true since diffractive optical elements, e.g. diffraction gratings and hologram lenses, are components through which rays have to be traced and they require an extension of Snell's law to take a ray through them. It may seem paradoxical that raytracing, which is a concept of geometrical optics, can be applied to diffraction gratings, which depend on physical optics for an explanation of their properties; it is found that the direction of a diffracted ray can be predicted by the extension of Snell's law mentioned above but the intensity of the radiation in the diffracted beam can only be found from physical optics arguments. Of course, the same applies to ordinary refraction, where the Fresnel formulae must be used to find the reflected and transmitted intensities, but whereas for ordinary refraction these effects can be neglected up to a certain stage in optical design because they are easily predictable from common experience, diffracted intensities vary greatly with the geometry and with the microstructure of the diffracting element; thus raytracing through diffractive elements predicts where a ray will go but not how intense it will be.

In this chapter we shall explain finite raytracing through toric surfaces, diffraction gratings and hologram lenses and mirrors.

5.1 Specification of toric surfaces

For optical purposes a toric is a surface generated by rotating a circle about an

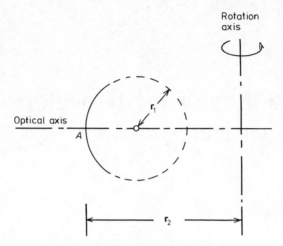

FIG. 5.1. Generating a ring toric surface

axis in its plane but not passing through its centre; it has two principal curvatures, corresponding to the radius of the generating circle and the distance from the rotation axis to the furthest point of the generating circle. Figure 5.1 shows a generating circle of radius r_1 and a rotation axis at a distance r_2 (which is greater than r_1) from a point A which will be on the equator of the toric. It can easily be seen that there is another way to generate a toric with the same two principal radii of curvature. This is shown in Fig. 5.2; the generating circle now has radius r_2 and the axis is distant r_1 from the point A on the equator.

If we take as the optical axis the axis of twofold symmetry formed by the line through A perpendicular to the rotation axis it can be seen that these two torics have the same paraxial properties (although as drawn one has the radius r_1 in the plane of the diagram and the other has it in a plane perpendicular to that of the diagram) but they differ away from the paraxial region. They are called respectively *ring* and *barrel* torics for obvious reasons and one or other might be used depending on the optical surfacing technology available; the ring form is mostly used for spectacle lens generation since more blanks can be blocked on a tool of given radius than for the barrel form.

To obtain the equation of the toric we set up the usual coordinate axes with origin at the pole of the toric (the point A in Figs. 5.1 and 5.2) and we use latitude and longitude angles θ and ϕ as in Fig. 5.3; we now denote the two radii of curvature by r_x and r_y as shown, and in what follows there will be no need to distinguish between ring and barrel torics.

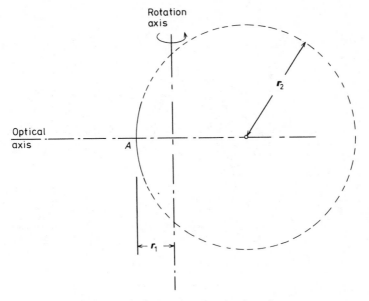

FIG. 5.2. Generating a barrel toric surface

The parametric equations of the toric are easily seen to be

$$\left.\begin{aligned}
x &= (r_x - r_y + r_y \cos \theta) \sin \phi \\
y &= r_y \sin \theta \\
z &= r_x - (r_x - r_y + r_y \cos \theta) \cos \phi
\end{aligned}\right\}. \tag{5.1}$$

Eliminating θ and ϕ and solving for z we obtain

$$z = r_x \pm [(r_x - r_y \pm (r_y^2 - y^2)^{\frac{1}{2}})^2 - x^2]^{\frac{1}{2}}. \tag{5.2}$$

The ambiguity signs correspond to four (possibly complex) values of z for given x and y, since the toric is a quartic surface. For optical purposes the appropriate solution is that for which z tends to zero with x and y, i.e. the result is

$$z = r_x - [(r_x - r_y + (r_y^2 - y^2)^{\frac{1}{2}})^2 - x^2]^{\frac{1}{2}}. \tag{5.3}$$

As in Section 4.3 this can be rearranged to conserve numerical precision in the form

$$z = \frac{F_t}{G_t\{G_t + (G_t^2 - c_x F_t)^{\frac{1}{2}}\}}, \tag{5.4}$$

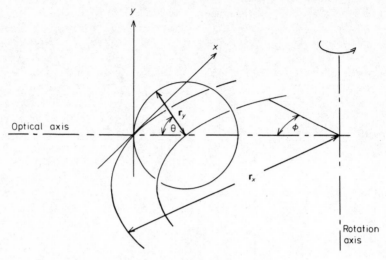

Fig. 5.3. Coordinates on a toric

where

$$F_t = 2G_t(c_x x^2 + c_y y^2) - c_x c_y^2(x^2 + y^2)y^2 \Bigg\} \\ G_t = 1 + (1 - c_y^2 y^2)^{\frac{1}{2}}$$ (5.5)

and c_x and c_y are the reciprocals of r_x and r_y, i.e. they are the two principal curvatures corresponding to the rotation axis and the generating circle.

The normal at (x, y, z) is given by

$$\mathbf{n} = (-\cos\theta \sin\phi, -\sin\theta, \cos\theta \cos\phi),$$ (5.6)

with

$$\theta = \arcsin c_y y \\ \phi = \arcsin \frac{G_t c_x x}{G_t - c_x c_y y^2} \Bigg\}$$ (5.7)

and the equation of the tangent plane at (x, y, z) is then obtained as in Chapter 4.

5.2 The transfer process for a toric

To find the intersection of a straight line with a toric surface requires the solution of a quartic or fourth degree algebraic equation; this can be done algebraically, but for our purposes it is simpler to use an iterative method, since an approximate solution for the required one of the four roots is always known. The method is similar to that described for general aspherics in Chapter 4. We start with the point P_1 at which the incoming ray meets the tangent plane at the axis, i.e. the plane $z = 0$, and find from eqn (5.4) the z-coordinate of the point P_1' on the toric which has the same x and y coordinates as P_1. Next we find the tangent plane at P_1' and take P_2 as the intersection of the ray with this plane. Then P_2' is the point on the toric with the same x and y coordinates as P_2 and so on until the coordinates of P_n and P_n' are the same to the required precision.

5.3 Refraction through a toric

Since we have the direction cosines (L, M, N) of the incident ray, the point of incidence (from Section 5.2) and the normal at the point of incidence (from Section 5.1), the refraction process is exactly as in Chapter 4.

5.4 Raytracing through diffraction gratings

It would be possible to treat the problem of raytracing through diffraction gratings as a special case of raytracing through holographic optical elements, particularly since many diffraction gratings are now made interferometrically, i.e. as, in effect, holographic optical elements. On balance, it seems better to treat diffraction gratings as a special case since they are generally thought of as having been ruled with a specified number of rulings per millimetre, rather than being a hologram. The main new thing to be taken account of is that there can be several diffracted rays for one incident ray, corresponding to a certain finite number of diffracted orders, and it is necessary to choose the appropriate order of diffraction; this will be denoted by m in what we may now call the diffraction algorithm, corresponding to refraction for lens surfaces.

A ruled concave grating has rulings corresponding to intersections of the (spherical) grating surface with equally spaced parallel planes; this is because of the method of manufacture but for no deeper reason. Thus in Fig. 5.4 with the usual coordinate system we can take the grating rulings as the intersections of the grating surface with the planes

$$y = 0, \pm\sigma, \pm 2\sigma, \ldots$$

where σ is known as the grating constant.

Consider a source point A of light of wavelength λ, as in Fig. 5.5; if the grating were a mirror a reflected wavefront Σ would be such that the optical path length APA' to a point A' on the wavefront would be constant as P

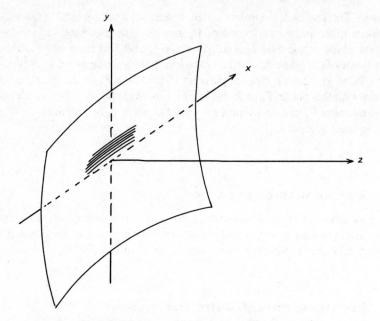

FIG. 5.4. Coordinate system for a concave diffraction grating

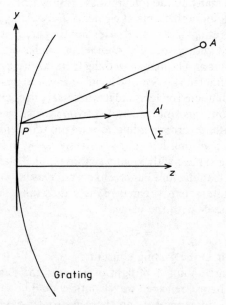

FIG. 5.5. Diffraction at a concave grating

explored the mirror. However, if Σ is actually a diffracted wavefront from the grating and the order of diffraction is m, then the optical path length from A to A′ must change by $m\lambda$ each time P crosses a ruling on the grating. Thus a quantity W must be constant as P explores the grating, where

$$W = AP + PA' + m\lambda y/\sigma. \tag{5.8}$$

The conditions that PA′ shall be the diffracted ray of order m corresponding to the incident ray AP are therefore

$$\frac{\partial W}{\partial x} = \frac{\partial W}{\partial y} = 0. \tag{5.9}$$

These two conditions will be shown to determine the direction of PA′, given the point P and the direction of AP.

Let A, P, and A′ have coordinates (ξ, η, ζ), (x, y, z) and (ξ', η', ζ') respectively. Then we have

$$(AP)^2 = (x - \xi)^2 + (y - \eta)^2 + (z - \zeta)^2 \tag{5.10}$$

with a similar result for PA′. Differentiating and substituting in eqns (5.9) we obtain

$$\left.\begin{aligned}
\frac{x-\xi}{AP} + \frac{z-\zeta}{AP} \cdot \frac{\partial z}{\partial x} + \frac{x-\xi'}{PA'} + \frac{z-\zeta'}{PA'} \cdot \frac{\partial z}{\partial x} &= 0 \\[2mm]
\frac{y-\eta}{AP} + \frac{z-\zeta}{AP} \cdot \frac{\partial z}{\partial y} + \frac{y-\eta'}{PA'} + \frac{z-\zeta'}{PA'} \cdot \frac{\partial z}{\partial y} + \frac{m\lambda}{\sigma} &= 0.
\end{aligned}\right\} \tag{5.11}$$

Now if the radius of curvature of the grating surface is R we have

$$\frac{\partial z}{\partial x} = \frac{x}{R-z}, \qquad \frac{\partial z}{\partial y} = \frac{y}{R-z} \tag{5.12}$$

and also for the direction cosines of the rays,

$$\left.\begin{aligned}
(L, M, N) &= (\xi - x, \eta - y, \zeta - z)/AP \\
(L', M', N') &= (\xi' - x, \eta' - y, \zeta' - z)/PA'
\end{aligned}\right\} \tag{5.13}$$

Substituting from eqns (5.12) and (5.13) into eqns (5.11) we obtain

$$\left.\begin{aligned}
L + L' + \frac{(N + N')x}{R-z} &= 0 \\[2mm]
M + M' + \frac{(N + N')y}{R-z} - \frac{m\lambda}{d} &= 0
\end{aligned}\right\} \tag{5.14}$$

or, with a slight rearrangement,

$$\frac{L + L'}{x} = \frac{M + M' - m\lambda/\sigma}{y} = \frac{N + N'}{z - R} = Q \tag{5.15}$$

where Q is to be determined. To find Q we rearrange eqns (5.15) as follows:

$$\left.\begin{array}{l} L' = xQ - L \\ M' = yQ - M + m\lambda/\sigma \\ N' = (z - R)Q - N. \end{array}\right\} \tag{5.16}$$

Then squaring and adding we obtain the quadratic equation for Q:

$$\{x^2 + y^2 + (z - R)^2\}Q^2 - 2\{Lx + (M - m\lambda/\sigma) + N(z - R)\}Q$$
$$- 2Mm\lambda/\sigma + m^2\lambda^2/\sigma^2 = 0. \tag{5.17}$$

To simplify this we note that the equation of the grating surface is

$$x^2 + y^2 + (z - R)^2 = R^2 \tag{5.18}$$

so that the normal at P is

$$\left(-\frac{x}{R}, -\frac{y}{R}, 1 - \frac{z}{R}\right) \tag{5.19}$$

and therefore, if I is the angle of incidence of the ray AP, we have

$$\cos I = -\frac{(Lx + My + N(z - R))}{R}. \tag{5.20}$$

Then substituting into eqn (5.17) we have as the equation for Q

$$Q^2 + \frac{2}{R}\left(\cos I + \frac{m\lambda y}{R\sigma}\right)Q - \frac{2Mm\lambda}{R^2\sigma} + \frac{m^2\lambda^2}{R^2\sigma^2} = 0. \tag{5.21}$$

The appropriate root can be chosen by noting that if the grating were simply a mirror Q would be given by $-(2\cos I)/R$; thus we have for Q

$$Q = -\frac{1}{R}\left\{\cos I + \frac{m\lambda y}{R\sigma} + \left[\left(\cos I + \frac{m\lambda y}{R\sigma}\right)^2 + \frac{2Mm\lambda}{\sigma} - \frac{m^2\lambda^2}{\sigma^2}\right]^{\frac{1}{2}}\right\} \tag{5.22}$$

and this may be used in eqns (5.16) for the diffraction algorithm, corresponding to the refraction process of Chapter 4. Clearly Q corresponds to the generalized power of eqn (4.14) and it reduces to that quantity when $m = 0$. The transfer operation is, of course, similar to that for refracting or reflecting surfaces, although there may be some slight extra complications if the system as a whole is not axisymmetric.

The diffraction algorithm can easily be modified for plane gratings by letting R tend to infinity. Equations (5.16) then take the following forms:

$$L' = -L, \qquad M' = -M + m\lambda/\sigma,$$
$$N' = -N + \cos I + \{\cos^2 I + 2Mm\lambda/\sigma - m^2\lambda^2/\sigma^2\}^{\frac{1}{2}}. \qquad\qquad (5.23)$$

5.5 Raytracing through holograms

Holographic optical elements are sometimes made on non-plane substrates; the forming or writing beams may be deliberately aberrated and the readout wavelength may be different from that used to write the hologram. We shall give a diffraction algorithm of sufficient generality to deal with all these cases. Just as for diffraction gratings, the raytracing process gives the directions of diffracted rays of chosen orders but, as explained in the introduction to this chapter, the process gives no information about the relative intensities of the diffracted beams. In particular the diffracted intensity from "thick" or volume holograms depends very strongly on fulfilment of the Bragg condition as well as the raytracing equations (see, for example, L. Solymar and D. J. Cooke. "Volume Holography and Volume Gratings", Academic Press 1981) but that is beyond the scope of the present book. In fact, we shall assume that the hologram can be represented by an indefinitely thin phase-changing layer of appropriate shape, as was tacitly assumed for diffraction gratings.

In order to accommodate all the possibilities mentioned above we assume that by means of suitable transfer algorithms rays from the forming beams and the readout beam have been traced to a certain point P on the hologram; the aim is then to determine the direction of the reconstructed or image beam at P. In Fig. 5.6 let n be a unit vector normal to the hologram at P, let r_0 and r_r be unit vectors along the two forming beams at P and let r_r' be a unit vector along the readout beam at P. We assume the hologram was formed in wavelength λ and the readout is to take place in wavelength λ'. Let r_0' be the required direction of the image beam at P. The principle of the method is as follows. All four beams are assumed to have plane phasefronts over a small enough region near P; the two forming beams then produce equi-spaced interference fringes which intersect the hologram surface to give a fringe pattern with a certain spacing and direction. The reconstructed beam must then be in a direction such that together with the readout beam it forms a fringe pattern on the hologram which either coincides with the first set or has a sub-multiple spacing.

We begin by finding the direction and spacing of the fringe pattern formed on the surface by the beams along r_0 and r_r. Let q be a unit vector parallel to

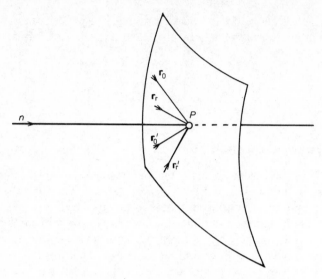

FIG. 5.6. Rays at a holographic optical element

these fringes; then q must be perpendicular to both n and $(r_0 - r_r)$, so that we must have

$$aq = n \wedge (r_0 - r_r) \tag{5.24}$$

where a is a scalar. By squaring eqn (5.24) and making use of elementary vector identities we find that a is given by

$$a = \{(r_0 - r_r)^2 - (n \cdot (r_0 - r_r))^2\}^{\frac{1}{2}}. \tag{5.25}$$

If P is displaced a small distance δr in the surface the optical path difference between the two forming beams increases by $\delta r \cdot (r_0 - r_r)$; then if δr is a displacement perpendicular to the direction of the fringes of magnitude one fringe, say σ, we must have

$$\delta r \cdot (r_0 - r_r) = \lambda \tag{5.26}$$

and also

$$\delta r = \sigma q \wedge n. \tag{5.27}$$

Substituting for δr in eqn (5.26) we have

$$\sigma(q \wedge n) \cdot (r_0 - r_r) = \lambda \tag{5.28}$$

and substituting for q from eqn (5.24) and again using vector identities we find for the fringe spacing

$$\sigma = \lambda/a \tag{5.29}$$

where a is given by eqn (5.25). Thus the hologram fringes at P are defined in direction by q from eqn (5.24) and in spacing by σ from eqn (5.29) and this completes the first half of the calculation.

On readout with wavelength λ' the vectors r'_0 and r'_r must satisfy the equation

$$n \wedge (r'_0 - r'_r) = a'q \tag{5.30}$$

where q is the same unit vector as was defined in eqn (5.24) but a' is a different scalar. However, the fringes corresponding to r'_0 and r'_r in wavelength λ' must either coincide with those already formed by r_0 and r_r in wavelength λ or they must be harmonics of the latter, so we must have

$$a' = m\lambda'/\sigma, \qquad m \text{ an integer} \tag{5.31}$$

and thus

$$n \wedge (r'_0 - r'_r) = m\lambda'q/\sigma. \tag{5.32}$$

Now comparing this with eqn (5.24) we have the vector raytracing equation for holograms,

$$n \wedge (r'_0 - r'_r) = \frac{m\lambda'}{\lambda} . n \wedge (r_0 - r_r) \tag{5.33}$$

where m is the order of diffraction; thus $m = 0$ gives the zero order or undeviated ray and $m = 1$ gives the ordinary first order diffracted ray. If in eqn (5.33) we take components of the vectors in a coordinate system with its z-axis along the local normal n, we obtain the two diffraction equations

$$\left. \begin{array}{l} L'_0 - L'_r = \dfrac{m\lambda'}{\lambda} (L_0 - L_r) \\[3mm] M'_0 - M'_r = \dfrac{m\lambda'}{\lambda} (M_0 - M_r). \end{array} \right\} \tag{5.34}$$

We notice that the direction and spacing of the local fringes do not appear explicitly in the final result. For numerical work with curved hologram substrates we can use eqns (5.34) by making a suitable coordinate rotation when the point of incidence has been found.

Alternatively we can avoid the coordinate transformation by a method due to R. W. Smith (private communication). By multiplying eqn (5.33) vectorially by n we obtain

$$n \wedge (n \wedge r'_0) = n \wedge U \tag{5.35}$$

where

$$U = n \wedge r'_r + \frac{m\lambda'}{\lambda} n \wedge (r_0 - r_r) \tag{5.36}$$

is known; that is,

$$(n \cdot r_0')n - r_0' = n \wedge U. \tag{5.37}$$

On squaring this equation and solving for $(n \cdot r_0')$ we obtain

$$(n \cdot r_0') = \{1 - U^2\}^{\frac{1}{2}} \tag{5.38}$$

since $(n \cdot U) = 0$. Then substituting back in eqn (5.37) we have as the solution

$$r_0' = n\{1 - U^2\}^{\frac{1}{2}} - n \wedge U. \tag{5.39}$$

Referring back to Section 5.5 it can be seen that the methods just described could be used for raytracing through diffraction gratings, since the grating rulings could be represented by holographic forming beams, both collimated and intersecting at an angle 2θ symmetrically about the normal, where the grating constant is given by

$$\sigma = \lambda/2 \sin \theta. \tag{5.40}$$

6 Optical Invariants

6.1 Introduction

We introduced the Lagrange invariant H in Section 3.3 as part of the structure of Gaussian optics but, as indicated at that point, it has a much wider significance; for example, H^2 represents in Gaussian approximation the total light flux through a given transverse section of an optical system and its invariance through a system is therefore a consequence of the conservation of energy.† Other invariants in an optical system can be found, involving finite rays, and, as with the Lagrange invariant, they have physical implications beyond the immediate geometrical optics. It also turns out that they provide neat and simple proofs of certain geometrical optics theorems which are otherwise not so directly proven. In this chapter we examine these invariants in detail.

6.2 Alternative forms of the Lagrange invariant

Figure 6.1 shows the configuration of an intermediate image, paraxial principal ray, paraxial ray from the centre of the object, etc. in an intermediate space of an optical system. We denote quantities belonging to the principal ray by a bar, so that its incidence height, convergence angle and intersection length are $\bar{h}$, $\bar{u}$ and $\bar{l}$ as shown. Let O be the axial object point, P the axial point of the refracting surface, C the centre of curvature of the refracting surface and P_1 the point of incidence of the paraxial ray through O which touches the rim of the pupil. Then in paraxial approximation hc is equal to the angle OCP_1 and thus in the triangle OCP_1 the external angle at P_1 is

† Reflection and absorption etc. are ignored.

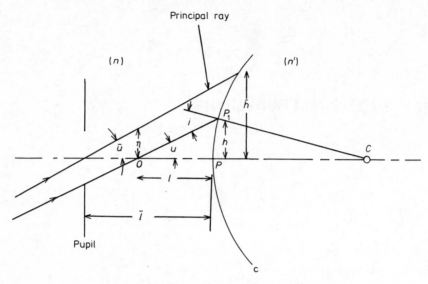

FIG. 6.1. Derivation of paraxial invariants

given by $hc + u$; this is the paraxial angle of incidence, which we denote by i, so we have found that

$$i = hc + u. \tag{6.1}$$

This equation serves to fix the sign convention for i, since that of u is fixed by its definition in eqn (3.19). Then from Snell's law we have

$$n'(hc + u') = n(hc + u). \tag{6.2}$$

Either side of this equation is denoted by A and this quantity is called the refraction invariant. It is essentially a paraxial quantity and it is, of course, only an invariant before and after refraction at a given surface, i.e. it is not true that $A_{+1} = A'$. It can be seen that eqn (6.2) could have been obtained by multiplying eqn (3.37) by h.

We define a refraction invariant $\bar{A}$ for the principal ray in the same way.

From Fig. 6.1 we have, with due attention to signs,

$$\eta = (l - \bar{l})\bar{u} \tag{6.3}$$

and so

$$H = nu\eta = nu\bar{u}(l - \bar{l}), \tag{6.4}$$

which is an alternative form of the Lagrange invariant.

Taking $u\bar{u}$ inside the bracket in eqn (6.4) we have also

$$H = n(u\bar{h} - \bar{u}h). \tag{6.5}$$

If we let u tend to zero in this result we recover the form of the Lagrange invariant for a star space (eqn (3.29)), since $\bar{u}$ is now the field angle.

Finally, using from eqn (6.2) $nu = A - nhc$ and $n\bar{u} = \bar{A} - nhc$, we have from eqn (6.5)

$$H = A\bar{h} - \bar{A}h. \tag{6.6}$$

The forms of the Lagrange invariant given in eqns (6.4–6.6) are used in primary aberration theory; they also have the merit that they make explicit the connection between the ray through the centre of the object and the ray through the centre of the pupil (principal ray) by the symmetry of the expressions in the barred and unbarred quantities.

6.3   The Seidel difference formulae

It may reasonably be conjectured that if the path of one paraxial ray through an optical system is known, then the path of any other ray can be found from it without the need for a new paraxial raytrace. For if the first ray is regarded as connecting a pair of axial conjugates, a second proposed ray may be regarded as a paraxial principal ray connecting a pair of conjugates at the edge of the field; if a segment of this second ray is given in any space, the Lagrange invariant then supplies a relationship between the two in all spaces, and this verifies the conjecture.

Several convenient formulae providing such a connection can be obtained from the results of Section 6.2; they are known as the Seidel difference formulae, after Ludwig Seidel, who gave them in a series of papers in the *Astronomische Nachrichten* (1856, Vol. 43) in which he obtained the formulae for the primary aberrations (Chapter 8). The main application of the Seidel difference formulae is in obtaining the data of a paraxial principal ray when the paraxial ray between the axial conjugates has been traced; this is needed for some forms of the primary aberration formulae.

Figure 6.2 shows the configuration of the two rays in an intermediate space; to fix our ideas we assume the known ray to be from the axial object point to the rim of the entrance pupil and the second to be the paraxial principal ray from the edge of the field. However, this is not essential and the relationships to be derived will apply to any pair of rays. The intermediate image and pupil are both shown as real for clarity and the bar notation is used for the quantities belonging to the principal ray, as in the previous section. Let

$$\bar{h} = hHE, \tag{6.7}$$

where H is the Lagrange invariant and E is a parameter of which the value is to be found at each surface; E is sometimes called the eccentricity, since it is a measure of $\bar{h}/h$, i.e. the relative displacement from the axis of the centre of the

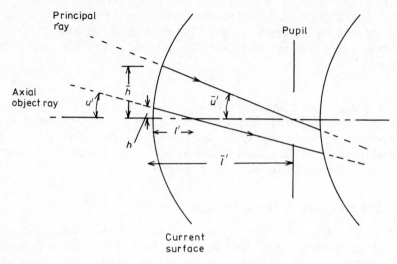

FIG. 6.2. Two rays in an intermediate space

image-forming pencil at the current refracting surface. If we suppose for the moment that E can be found we can obtain all the other quantities relating to the principal ray as follows. First from eqn (6.6) we have:

$$\bar{A} = \frac{A\bar{h} - H}{h},$$
(6.8)

or, substituting for $\bar{h}/h$,

$$\bar{A} = -\frac{H}{h}(1 - AhE).$$
(6.9)

Similarly from eqn (6.5),

$$n\bar{u} = \frac{nu\bar{h} - H}{h}$$
(6.10)

or,

$$n\bar{u} = -\frac{H}{h}(1 - nuhE).$$
(6.11)

Thus provided the eccentricity E is known at each surface we can find all the data of the principal ray from eqns (6.7), (6.9) and (6.11). We have for any two successive surfaces, from eqn (6.7),

$$H(E_{+1} - E) = \frac{\bar{h}_{+1}}{h_{+1}} - \frac{\bar{h}}{h}$$
(6.12)

$$= \frac{\bar{h}_{+1}h - \bar{h}h_{+1}}{h_{+1}h}.$$
(6.13)

Using the results $h_{+1} = -l_{+1}u' = -(l' - d')u'$ and similarly for the barred quantities we obtain:

$$H(E_{+1} - E) = \frac{(\bar{l}' - d')\bar{u}' . l'u' - (l' - d')u' . \bar{l}'\bar{u}'}{h_{+1}h},$$ (6.14)

$$= -\frac{\bar{u}'u'(l' - \bar{l}')d'}{h_{+1}h}.$$ (6.15)

Thus on using eqn (6.4) we obtain

$$E_{+1} - E = \frac{-d'}{n'hh_{+1}}.$$ (6.16)

Thus the increment in the eccentricity on going from one surface to the next is obtained from eqn (6.16) in terms of the separation between the surfaces and the incidence heights of the ray from the axial object point at the surfaces. Now let the actual aperture stop be at a known position between surfaces k and $k + 1$, as in Fig. 6.3. We can treat the stop plane as a ghost surface and

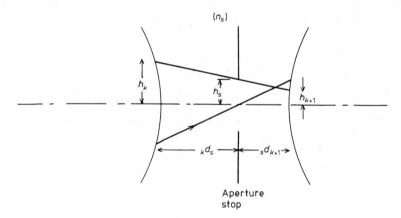

FIG. 6.3. Computing the eccentricity parameter

find the incidence height of the ray from the axial object point there, say h_s. From the definition in eqn (6.7) it can be seen that the eccentricity is identically zero at the stop, i.e. $E_s = 0$. Then we have:

$$E_{k+1} = \frac{-{}_sd_{k+1}}{n_sh_sh_{k+1}}$$ (6.17)

and also

$$E_k = \frac{_k d_s}{n_s h_k h_s} \qquad (6.18)$$

The values of E at the other surfaces are then found by working in both directions from the stop according to eqn (6.16).

The above discussion referred to the second ray, of which the parameters were to be found, as the principal ray, but it can be any other paraxial ray. For example, if the ray which has been traced is from an axial conjugate at a finite distance we can find the parameters of a ray from the axial object at infinity and hence the equivalent focal length. To do this we take the position of the "stop" to be at infinity in object space and then the second ray will pass through F'.

6.4 The skew invariant

We now come to an invariant for finite rays. Let r in Fig. 6.4 be a skew ray in object space and let OP be the line which is perpendicular to the axis and to the ray r; clearly OP is the shortest distance between the ray and the axis; we denote it by S and we also denote the angle between r and the axis by

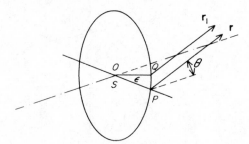

FIG. 6.4. The skew invariant

θ. If the complete ray with all its segments in all spaces is rotated through an angle ε about the axis it is still a physically possible ray r_1 on account of the cylindrical symmetry. Let ε be small, let Q be the new position of P after rotation and let R be the foot of the perpendicular from P to the new ray r_1. There will be a similar diagram for the rays r and r_1 in image space. Since r_1 was obtained from r by a rigid rotation we have

$$[PP'] = [QQ'] \qquad (6.19)$$

and since PR is perpendicular to both rays we have

$$[PP'] = [RR'].\qquad(6.20)$$

Therefore

$$[R'Q'] = [RQ],\qquad(6.21)$$

or

$$(n'S' \sin \theta')\varepsilon = (nS \sin \theta)\varepsilon.\qquad(6.22)$$

Thus we have the result that the quantity

$$\mathcal{H} = nS \sin \theta\qquad(6.23)$$

is invariant through the system. It is not known who first discovered the *skew invariant* but many authors have independently rediscovered it; it was given, for example, by T. Smith in 1921. However, little use seems to have been made of it in deriving general properties of optical systems until a note by the present author (*Optica Acta* **15**, 621–3 (1968)).

It is useful to express the skew invariant in terms of the raytracing quantities used in Chapter 4. Let (x, y, z) be any point on a skew ray of which the direction cosines are (L, M, N); the equations of the ray are

$$x' = x + \frac{L}{N}(z' - z),\qquad y' = y + \frac{M}{N}(z' - z),\qquad(6.24)$$

(x', y', z') being the coordinates of the current point. The square of the perpendicular distance of (x', y', z') from the z-axis, i.e. the optical axis, is therefore

$$x^2 + y^2 + 2\frac{(Lx + My)}{N}(z' - z) + \frac{L^2 + M^2}{N^2}(z' - z)^2.\qquad(6.25)$$

This expression is stationary when the derivative with respect to z' is zero, i.e. when

$$z' - z = -\frac{N(Lx + My)}{L^2 + M^2};\qquad(6.26)$$

substituting this value in the expression (5.25) we find

$$S^2 = \frac{M^2x^2 - 2LMxy + L^2y^2}{L^2 + M^2}\qquad(6.27)$$

and therefore since $\sin \theta = \sqrt{1 - N^2}$ we have,

$$\mathcal{H} = n(Mx - Ly).\qquad(6.28)$$

This result could be used as an arithmetical check on skew raytraces when, for example, x and y could be taken as coordinates of the point of incidence. In fact it has much wider applications, as will be seen in the next section, since x and y can refer to any point on the ray.

6.5 Some applications of the skew invariant

Off-axis object points are usually put in the y–z plane by convention, i.e. they have zero x coordinates. Let a skew ray of a pencil from such an off-axis object point P meet the y–z plane again in image space at P_1; the point P_1 has been called by M. Herzberger the diapoint corresponding to P *via* this ray. It follows immediately from eqn (6.28) that if η and η' are the distances of P and P_1 from the axis then

$$n'L'\eta' = nL\eta. \tag{6.29}$$

This relation obviously applies also to all diapoints in the intermediate image spaces.

Now suppose the skew ray to be very close to the y–z plane, as in Fig. 6.5, and let a ray in the y–z plane be drawn from P to P_1 having the same values of

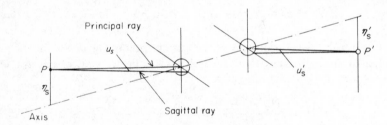

FIG. 6.5. Close sagittal rays

N, the z-direction cosine; this latter ray can then be regarded as a principal ray from P and the skew ray bears much the same relation to it as does a paraxial ray to the optical axis. If then we denote the quasi-paraxial angle between the skew ray and the principal ray by u_s the skew invariant formula gives:

$$n'u_s'\eta_s' = nu_s\eta_s. \tag{6.30}$$

The skew ray lies in each space in a plane through the principal ray and at right angles to the y–z plane; these planes are said to be sagittal planes, the skew ray is a (close) sagittal ray and the foci P and P_1 are sagittal foci (see Section 9.8).

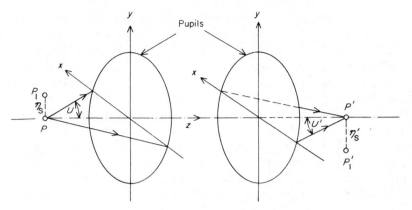

FIG. 6.6. The optical sine theorem

Finally consider two axial points P and P′ which are conjugates for a ray with finite convergence angles U and U'; two such rays, lying in, say, the x–z plane, are indicated in Fig. 6.6. Let P now be displaced a *small* distance η_s to P_1 in a direction perpendicular to the x–z plane and let the corresponding point in image space be P_1'. The imagery between P_1 and P_1' is by rays lying close to the x–z plane, i.e. rays which are at right angles to the segments PP_1 and $P'P_1'$, so that these segments are the shortest distances in object and image spaces between these rays and the z-axis. The skew invariant formula then states

$$n'\eta_s' \sin U' = n\eta_s \sin U, \tag{6.31}$$

a result sometimes called the *optical sine theorem*. This result gives a magnification for small distances imaged by rays at large angles, whereas eqn (6.30) is for large distances with rays at small angles to the meridian plane. It is important to remember that the segments PP_1 and $P'P_1'$ are perpendicular to the plane containing the rays since it is only in this case that eqn (6.31) is valid; if the segments were taken in the same plane as the rays there would be no generally true simple result of this kind, but on the other hand the *Abbe sine condition* for the absence of coma can then be applied; this has a rather similar-looking mathematical expression but it is a condition which holds only when there is no coma, whereas eqn (6.31) is true for all symmetrical optical systems.

6.6 The generalized Lagrange invariant

It was noted in Section 6.1 that H^2 is, in Gaussian approximation, proportional to the light flux emitted by the area of the axial object in the solid angle collected by the entrance pupil; it corresponds to the *étendue* in spectroscopic instrumentation. This suggests the question: is there a corresponding invariant, not in Gaussian approximation, which represents the light flux collected from an element at a finite distance from the axis? Such an invariant exists and although it has not so far been very important in optical design we shall give it here because of its extreme generality.

The result is applicable to optical systems without any axis of symmetry; the derivation we shall give depends on the point characteristic function V of Hamilton, which is formally discussed in Section 7.2. This function is the optical path length between any two points P and P′, in object and image space respectively, and its principal property, proved in Section 7.2, is that its derivatives with respect to the coordinates of P or P′ give the corresponding

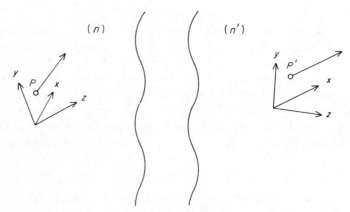

FIG. 6.7. Coordinates for the generalized Lagrange invariant

direction cosines of the ray segment through these points. We take rectangular axes arbitrarily oriented in object and image space as in Fig. 6.7 and we restrict P and P′ to lie on the x–y and x'–y' planes respectively. Then for the ray through P and P′ we have, from Section 7.2,

$$\left.\begin{array}{ll} -nL = \dfrac{\partial V}{\partial x} & n'L' = \dfrac{\partial V}{\partial x'} \\[4mm] -nM = \dfrac{\partial V}{\partial y} & n'M' = \dfrac{\partial V}{\partial y'} \end{array}\right\} \tag{6.32}$$

It is not supposed that P and P′ are in any special object–image relationship.

Now consider small displacements dx and dy of P and small changes dL and dM of the direction cosines of the ray through P. Then the generalized

Lagrange invariant is the quantity $n^2\, dx\, dy\, dL\, dM$, so that we have

$$dx\, dy\, dp\, dq = dx'\, dy'\, dp'\, dq', \qquad (6.33)$$

where we have put $p = nL$, $q = nM$, $p' = n'L'$, $q' = n'M'$.

The proof of this result is most easily carried out by making use of some properties of matrices.

By differentiating eqns (6.32) we have

$$\left.\begin{aligned}
dp &= -V_{xx'}\, dx' - V_{xy'}\, dy' - V_{xx}\, dx - V_{xy}\, dy \\
dq &= -V_{yx'}\, dx' - V_{yy'}\, dy' - V_{yx}\, dx - V_{yy}\, dy \\
dp' &= V_{x'x'}\, dx' + V_{x'y'}\, dy' + V_{x'x}\, dx + V_{x'y}\, dy \\
dq' &= V_{y'x'}\, dx' + V_{y'y'}\, dy' + V_{y'x}\, dx + V_{y'y}\, dy
\end{aligned}\right\} \qquad (6.34)$$

where $V_{xx} = \partial^2 V/\partial x^2$, $V_{xy} = \partial^2 V/\partial x\, \partial y$, etc.

In matrix notation we can express eqns (6.34) as

$$\begin{pmatrix} V_{xx'} & V_{xy'} & \cdot & \cdot \\ V_{yx'} & V_{yy'} & \cdot & \cdot \\ V_{x'x'} & V_{x'y'} & -1 & \cdot \\ V_{y'x'} & V_{y'y'} & \cdot & -1 \end{pmatrix} \begin{pmatrix} dx' \\ dy' \\ dp' \\ dq' \end{pmatrix} = \begin{pmatrix} -V_{xx} & -V_{xy} & -1 & \cdot \\ -V_{yx} & -V_{yy} & \cdot & -1 \\ -V_{x'x} & -V_{x'y} & \cdot & \cdot \\ -V_{y'x} & -V_{y'y} & \cdot & \cdot \end{pmatrix} \begin{pmatrix} dx \\ dy \\ dp \\ dq \end{pmatrix}$$

$$(6.35)$$

Denoting the two matrices by A and B respectively and the column vectors of differentials by M' and M, this equation may be written

$$AM' = BM, \qquad (6.36)$$

so that

$$M' = A^{-1}BM. \qquad (6.37)$$

This matrix equation can be expanded to

$$dx' = \frac{\partial x'}{\partial x}\, dx + \frac{\partial x'}{\partial y}\, dy + \frac{\partial x'}{\partial p}\, dp + \frac{\partial x'}{\partial q}\, dq$$

and three similar equations, so that

$$\det (A^{-1}B) = \frac{\partial(x', y', p', q')}{\partial(x, y, p, q)}. \qquad (6.38)$$

But this is simply the Jacobian which transforms the differential hypervolume element $dx\, dy\, dp\, dq$, i.e. we must have

$$dx'\, dy'\, dp'\, dq' = \frac{\partial(x', y', p', q')}{\partial(x, y, p, q)}\, dx\, dy\, dp\, dq, \qquad (6.39)$$

so that eqn (6.33) will be proved if we can show that

$$\det(A^{-1}B) = 1. \tag{6.40}$$

But

$$\det(A) = \begin{vmatrix} V_{xx'} & V_{xy'} \\ V_{yx'} & V_{yy'} \end{vmatrix} = Q, \quad \text{say} \tag{6.41}$$

and det (B) has the same value, remembering that $V_{xy'} = V_{y'x}$, etc. Also the determinant of the product of two matrices is the product of their determinants, from which it follows that $\det(A^{-1}) = Q^{-1}$. Thus eqn (6.40) is proved and so eqn (6.33) is true.

This result has been given by several authors† and it has been pointed out that it is the geometrical optics analogue of Liouville's theorem in statistical mechanics; in the present application the phase space is four-dimensional, the optical direction cosines p and q are analogues of generalized momenta and the spatial coordinates x and y are like generalized position coordinates. The z-axis corresponds to the evolution of time. The analogy can be developed from Fermat's principle; this would usually be expressed by saying that the optical path length W between two fixed points is stationary along a physically possible ray path and W is given by

$$W = \int n \, ds. \tag{6.42}$$

However if W is written in the form

$$W = \int n(x, y, z) \left[1 + \left(\frac{dx}{dz} \right)^2 + \left(\frac{dy}{dz} \right)^2 \right]^{\frac{1}{2}} dz \tag{6.43}$$

the integrand is of the form of the Lagrangian L of generalized mechanics, with z as the time axis. This formulation is a convenient starting point for developing the geometrical optics of media of continuously varying refractive index, e.g. graded index fibres. Also in this analogy the skew invariant of Section 6.4 corresponds to conservation of angular momentum with no frictional forces.

Returning to geometrical optics, the significance of the generalized Lagrange invariant is that in the geometrical optics model the light flux emitted by an element $dx \, dy$ of a Lambertian source in the x–y plane over a range of solid angle specified by $dL \, dM$ is constant through the optical system, apart from reflection and absorption losses, assuming that all the rays of this infinitesimal pencil do get through the system. Thus the volume element $dx \, dy \, dp \, dq$ in phase space which represents this pencil of rays changes in shape as it travels through the system but its volume is constant. The analogy

† For example, G. Toraldo di Francia, *J. Opt. Soc. Am.* **40**, 600 (1950); O. N. Stavroudis, *J. Res. nat. Bur. Stand.* **63B**, 31–42 (1959); and D. Marcuse, "Light Transmission Optics", Van Nostrand–Reinhold, New York (1972).

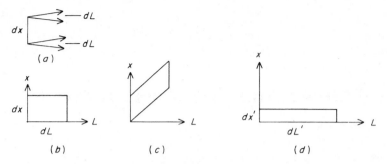

FIG. 6.8. The generalized Lagrange invariant traced in phase space

with the paraxial Lagrange invariant H of Section 3.3 is obvious but the generalized Lagrange invariant applies to any optical system, symmetrical or not and image-forming or not. The directions of the coordinate axes used in defining the invariant can be chosen arbitrarily in the different spaces but in any one space the direction cosine elements must be referred to the same axes as the length elements dx and dy. It is, of course, assumed that the differentiations carried out in eqns (6.32) and (6.34) are permissible; this is true for optical systems with smooth surfaces provided singularities such as rays being tangent to refracting surfaces or suffering total internal reflection are avoided. It is also to be understood that if the characteristic function V is multivalued between the points P and P' in Fig. 6.7 the invariant applies separately along each ray between these points.

The meaning of the generalized Lagrange invariant can be illustrated by considering a pencil of rays in one plane only, so that it becomes $n\,dx\,dL$, as in Fig. 6.8a; the corresponding "volume" in phase space is as in Fig. 6.8b and after propagating some distance in the same medium the rays will have spread and the volume appears as in Fig. 6.8c. Finally after refraction the volume might appear at the conjugate plane to the original element as in Fig. 6.8d; the area of the figure multiplied by the refractive index of the medium is constant.

The ordinary Lagrange invariant H, the optical sine theorem (eqn (6.31)) and other results could be deduced as special cases from the generalized Lagrange invariant. However, it is a somewhat abstract concept for ordinary glass optics, although the particle optics version is of great practical use; it therefore seems more appropriate to use special derivations for these theorems.

7 Monochromatic Aberrations

7.1 Introduction: definitions of aberration

We have already met the general concept of aberration in an optical system as the failure of the system to conform to the mode of ideal image formation described in Chapter 1. If P is any object point and P′ is its Gaussian image then all rays from P, i.e. all those transmitted by the aperture stop, should pass through P′; alternatively the wavefronts from P should, in image space, be portions of spheres centred on P′. These conditions are equivalent, according to Chapter 2, and aberrations appear as non-fulfilment of either condition. It is possible that the rays and wavefronts from P might all converge to a point P″ in image space which is not P′, the Gaussian image of P, but which is near P′. This leads to a very useful distinction between point-imaging aberrations, in which the rays in image space are not concurrent, and aberrations of image shape, where each object point forms a true image point but there is not the correct similarity between object and image shapes. The latter aberrations are, of course, field curvature and distortion while the former are spherical aberration, coma, astigmatism and higher order compounds of complex form.

The finite raytracing process described in Chapter 4 will clearly give unlimited information about aberrations as ray deviations, provided enough rays are traced, and this is a relatively simple matter if a fast computer is available; it is also fairly simple to supplement the raytracing procedure to obtain the wavefront shape errors directly, as will be shown in Section 7.2. Thus in principle we have already solved the problem of determining the aberrations of a given optical system. However, the inverse problem, that of obtaining a system to have given aberrations, i.e. zero aberrations to within certain tolerances, is the basic problem of optical design and in order to attempt to solve this we need to have some notions of how aberrations arise, or what kinds of aberrations are produced by given kinds of optical elements. This leads to a more detailed classification of aberrations and to systems of formulae which give explicitly the more important aberrations in terms of the structure of the system, the primary aberration formulae of Chapter 8.

In the present chapter we shall examine the relationship between the two main definitions of aberration and we shall introduce a classification of aberrations with formulae and nomenclature suitable for the purposes of optical design. These relationships between the definitions are best obtained by introducing the *characteristic function* of Hamilton, a remarkable concept which can be used to obtain several useful theoretical results although in its original form it does not have much direct application in optical design. We restrict our discussion here to monochromatic aberrations, i.e. we do not yet consider effects due to change of refractive index with wavelength. Hamilton's work was published in the 1830s and can be found in Vol. 1 of his collected papers ("The Mathematical Papers of Sir W. R. Hamilton" (A. W. Conway and J. L. Synge, eds), Cambridge University Press, 1931).

7.2 Wavefront aberrations, transverse ray aberrations and characteristic functions

Let P and P' be points in the object and image spaces of an optical system which are not in general conjugates, so that usually only one ray can be found to pass through P and P'. Hamilton's point characteristic function V is defined as the optical path length along this unique ray from P to P'. Thus it is a function of the coordinates of P and P' and may be written $V(a, a')$ where a and a' are position vectors of P and P'.† The functional form of V depends, of course, on the optical system and it is only possible to compute it numerically, i.e. it cannot be obtained by a closed formula, except in trivial cases. Ignoring this difficulty, suppose that V is known for a given system. Let r' be a unit vector on the ray through P' (Fig. 7.1) and let Σ' be the wavefront from P through P'. If P'' is a neighbouring point with position vector $a' + \delta a'$ we have, regarding V as a function of a' only for the moment

$$V(a' + \delta a', a) - V(a', a) = \delta a' \cdot grad\ V. \tag{7.1}$$

But the left-hand side of this equation is equal to the optical path length between the wavefronts Σ' and Σ'' through P' and P'' respectively, i.e. it is

† If P' is near the Gaussian image of P it may happen that two or more rays from P pass through P' and the optical paths may not be equal. The function V is then multivalued.

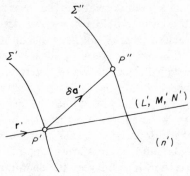

FIG. 7.1. The characteristic function and the ray components

$n'\,\delta a' \cdot r'$, where n' is the refractive index of the medium. Thus we have

$$n'\,\delta a' \cdot r' = \delta a' \cdot \mathbf{grad}\ V \qquad (7.2)$$

and, since $\delta a'$ is an arbitrary increment, we have $n'r' = \mathbf{grad}\ V$ or

$$\left. \begin{aligned} n'L' &= \frac{\partial V}{\partial x'} \\[2mm] n'M' &= \frac{\partial V}{\partial y'} \\[2mm] n'N' &= \frac{\partial V}{\partial z'} \end{aligned} \right\} \qquad (7.3)$$

where, as usual, L', M', N' are the direction cosines of the ray. In the same way we obtain in the object space,

$$\left. \begin{aligned} nL &= -\frac{\partial V}{\partial x} \\[2mm] nM &= -\frac{\partial V}{\partial y} \\[2mm] nN &= -\frac{\partial V}{\partial z} \end{aligned} \right\} \qquad (7.4)$$

Thus the point characteristic function would, if known, give completely the object–image correspondence of rays. In fact it can only be obtained by extensive numerical ray tracing so that this property is of no direct practical value. Its value is as a foundation for the general theory.

Another, closely related, characteristic function is called the *eikonal* and is denoted by E. This function was also defined and used by Hamilton, but it was discovered independently and given the name eikonal by H. Bruns in 1895.

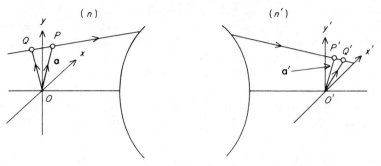

FIG. 7.2. The eikonal

Let O and O' in Fig. 7.2 be origins in object and image space, not necessarily conjugates, and let a ray meet the planes perpendicular to the axis through O and O' in P and P'; let OP be the vector a and O'P' the vector a'; finally let the perpendiculars from O and O' meet the ray at Q and Q'. The eikonal E is defined as the optical path length from Q to Q', regarded as a function of the direction cosines of the ray segments. Thus in terms of the characteristic function we have

$$E(r, r') = V(a, a') - n'a' \cdot r' + na \cdot r. \tag{7.5}$$

By differentiating this equation with respect to L' we have

$$\frac{\partial E}{\partial L'} = \frac{\partial V}{\partial x'} \frac{\partial x'}{\partial L'} + \frac{\partial V}{\partial y'} \frac{\partial y'}{\partial L'} - n'x' - n'L'\frac{\partial x'}{\partial L'} - n'M'\frac{\partial y'}{\partial L'}; \tag{7.6}$$

on using the first two of eqns (7.3) we have

$$\frac{\partial E}{\partial L'} = -n'x' \tag{7.7}$$

and similarly

$$\frac{\partial E}{\partial M'} = -n'y'. \tag{7.8}$$

Again there are corresponding equations in object space and thus the eikonal would also give completely the properties of the system if it could be computed, since eqns (7.7) and (7.8) would yield the transverse ray aberrations as defined below.

We can now define wavefront aberration and, through the characteristic functions, relate it directly to ray aberration. Let O in Fig. 7.3 be the centre of the exit pupil of a symmetrical optical system and let OP'_0 be a finite principal ray from an object point P, not shown; we take the usual x-, y-axes in the pupil and ξ-, η-axes in the image plane, but in the present case neither pupil nor image plane is necessarily the Gaussian plane, although it may be convenient to suppose them to be so in the first instance. The point P is in the plane containing the y- and z-axes; there is no loss of generality in this on account of the axial symmetry of the system. Let Σ' be the wavefront of the pencil from P which passes through O and let S be a *reference sphere* with centre P_0' and radius $P_0'O$. Let another ray r of the pencil from P meet S and Σ' in Q_0 and Q respectively and let it meet the image plane at P'; the co-ordinates of Q_0, P' and P_0' will be (x, y, z), (ξ, η) and $(0, \eta_0)$ respectively and

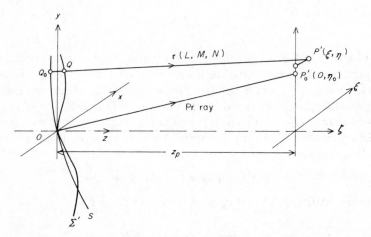

FIG. 7.3. Wavefront aberration and transverse ray aberration

the direction cosines of the ray r will be (L, M, N). The wavefront aberration is then defined as the optical path length from Q_0 to Q, i.e. it is nQ_0Q, where n is the refractive index of the space in question. We use the symbol W for wavefront aberration and in the present situation it is a function of the position of Q_0, a point on the reference sphere; thus we write it as $W(x, y)$, since z is determined by the condition that Q_0 lies on the reference sphere. Clearly the function $W(x, y)$ expresses the deformation of the wavefront from the ideal spherical shape and it is therefore a suitable function to specify the aberration according to the second mode described in Section 7.1.

We can also use Fig. 7.3 to frame a definition of ray aberration, according to the first mode of 7.1. The fact that P' does not coincide with P_0' constitutes aberration and we can use the two components of the displacement $P_0'P'$ to specify the aberrations; these are ξ, $\eta - \eta_0$ and they are called the components of transverse ray aberration. Again they are functions of x and y. We denote them by $\delta\xi$ and $\delta\eta$.

Both wavefront aberration and transverse ray aberration also depend of course on the position of the object point P, but we are not at present concerned with this.

We can now use the characteristic function to obtain the connection between wavefront aberration and transverse ray aberration. The optical path length from P to Q is $V(P, Q)$, where V is the point characteristic function as defined above, so that

$$
\begin{aligned}
W(x, y) &= V(P, Q) - V(P, Q_0) \\
&= V(P, O) - V(P, Q_0),
\end{aligned}
\tag{7.9}
$$

since Q and O are on the same wavefront and are therefore at the same optical distance from P. Differentiating this equation we have

$$
\left.
\begin{aligned}
\frac{\partial W}{\partial x} &= -\frac{\partial V}{\partial x} - \frac{\partial V}{\partial z} \cdot \frac{\partial z}{\partial x} \\
\frac{\partial W}{\partial y} &= -\frac{\partial V}{\partial y} - \frac{\partial V}{\partial z} \cdot \frac{\partial z}{\partial y}
\end{aligned}
\right\}
\tag{7.10}
$$

The equation of the reference sphere is

$$
x^2 + y^2 + z^2 - 2\eta_0 y - 2z_p z = 0,
\tag{7.11}
$$

where z_p is the distance of the image plane from the pupil. If we substitute the values obtained from this for $\partial z/\partial x$ and $\partial z/\partial y$ into eqn (7.10) and if we use eqn (7.3) we have:

$$
\left.
\begin{aligned}
\frac{1}{n}\frac{\partial W}{\partial x} &= -\left(L + \frac{Nx}{z_p - z}\right) \\
\frac{1}{n}\frac{\partial W}{\partial y} &= -\left(M + \frac{N(y - \eta_0)}{z_p - z}\right)
\end{aligned}
\right\}.
\tag{7.12}
$$

But

$$
(L, M, N) = \frac{(\xi - x, \eta - y, z_p - z)}{Q_0 P'}
\tag{7.13}
$$

and substituting these values in eqn (7.12) we have

$$
\left.
\begin{aligned}
\delta\xi &= -\frac{Q_0 P'}{n}\frac{\partial W}{\partial x} \\
\delta\eta &= -\frac{Q_0 P'}{n}\frac{\partial W}{\partial y}
\end{aligned}
\right\}.
\tag{7.14}
$$

These equations relating the wavefront aberration to the transverse ray aberrations are exact, but they contain the unknown distance $Q_0 P'$. For all practical purposes we can replace Q_0P' by R, the radius of the reference sphere, and we then have

$$
\left.
\begin{aligned}
\delta\xi &= -\frac{R}{n}\frac{\partial W}{\partial x} \\[2ex]
\delta\eta &= -\frac{R}{n}\frac{\partial W}{\partial y}
\end{aligned}
\right\}
\tag{7.15}
$$

as the expressions for transverse ray aberration components in terms of the wavefront aberration. It should be noted that the coordinates x and y are those of Q_0, the point of intersection of the ray with the reference sphere, not the pupil plane; under this condition eqns (7.14) are true for any field angle.

It happens more frequently that the transverse ray aberrations are known from raytracing and we wish to calculate the wavefront aberration; thus we know $\delta\xi$ and $\delta\eta$ as functions of x and y for some locus across the reference sphere from say A to B; then from eqns (7.15) we have

$$
\frac{R}{n}(W_B - W_A) = -\int_A^B \{\delta\xi\,\mathrm{d}x + \delta\eta\,\mathrm{d}y\},
\tag{7.16}
$$

where the path of integration is from A to B.

7.3 The effect of a shift of the centre of the reference sphere on the aberrations

In Section 7.2 we discussed the relationship between transverse ray aberrations and wavefront aberrations with a particular choice of the centre of the reference sphere, namely the intersection of the principal ray with the image plane. The choice was made in order to fix the ideas for the discussion in hand, but other choices are sometimes desirable. More generally, we need to know the effect on the aberrations of a small shift in the centre of the reference sphere; this will enable us to choose as the nominal image point that centre with respect to which the aberrations are minimal, according to some agreed criterion.

Suppose that, in the notation of Section 7.2, the point P_0' is shifted to P_1', a position with coordinates $(\delta\xi_0, \eta_0 + \delta\eta_0, \delta\zeta_0)$ and let the new reference sphere still pass through O, the centre of the pupil. If the resulting increment of wavefront aberration is δW as in Fig. 7.4, then by differentiating the equation of the reference sphere (eqn (7.11)) we have

$$
(xL + yM + zN)\frac{\delta W}{n} = -x\,\delta\xi_0 - y\,\delta\eta_0 + \eta_0 M\frac{\delta W}{n}
$$

$$
- z\,\delta\zeta_0 + z_pN\frac{\delta W}{n}.
\tag{7.17}
$$

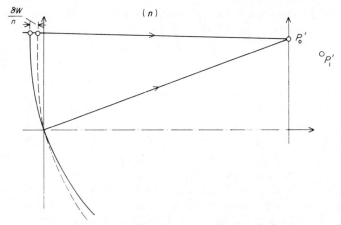

FIG. 7.4. The effect of shifting the centre of the reference sphere

since $\delta x = -L\,\delta W/n$, etc.; for this differentiation it was necessary to put in eqn (7.11) a notional term $-2\xi_0 x$ and to put $\xi_0 = 0$ *after* differentiation. Thus we have

$$\begin{aligned}
\delta W &= -\frac{n(x\,\delta\xi_0 + y\,\delta\eta_0 + z\,\delta\zeta_0)}{Lx + M(y - \eta_0) + N(z - z_p)} \\
&= \frac{n}{R}(x\,\delta\xi_0 + y\,\delta\eta_0 + z\,\delta\zeta_0).
\end{aligned}$$

(7.18)

Thus we have obtained the change in wavefront aberration due to the shift in the centre of the reference sphere.

For the transverse ray aberrations it is obvious that we have for the increments in $\delta\xi$ and $\delta\eta$,

$$\left.\begin{aligned}
\delta(\delta\xi) &= \frac{x}{R}\,\delta\zeta_0 - \delta\xi_0 \\
\delta(\delta\eta) &= \frac{y}{R}\,\delta\zeta_0 - \delta\eta_0
\end{aligned}\right\}.$$

(7.19)

7.4 Physical significance of the wavefront aberration

Following on the discussion of Section 7.3 it might be asked: What is the effect on the wavefront aberration of a change in the radius R of the reference sphere (but keeping its centre fixed)? It can be shown that if the change in R

is δR then the change in W is of order $\theta^2\,\delta R$, where θ is the small angle $P_0'P'/R$, i.e. the angular ray aberration. This quantity is generally very much smaller than the actual wavefront aberration. Nevertheless it may be felt that it is unsatisfactory to have defined a measure of aberration of which the magnitude depends on the apparently arbitrary choice of the radius of the reference sphere, particularly since there is, of course, no such effect for the transverse ray aberrations nor for the characteristic functions of Section 7.2. This disadvantage of the wavefront aberration does not, however, seriously affect its main utility, which is to provide a measure of aberration with direct physical significance for the formation of an optical image. To see how this comes about we have to refer to some concepts of physical optics, anticipating the discussion in Chapter 13. The image of a point object formed by an aberration-free system is not, of course, a point but a diffraction pattern of which the scale depends on the wavelength and the convergence angle of the image-forming pencil, the well-known Airy pattern. In the presence of aberrations this pattern becomes less sharp and this effect can be regarded as due to the fact that the light disturbances from different parts of the pupil do not arrive at the geometrical image point (the centre of the reference sphere) with the same phase; the phase of the disturbance from the point (x, y) in the pupil is in fact $(2\pi/\lambda)W(x, y)$ where W is the wavefront aberration as defined in Section 7.2, and this quantity enters directly into the calculation of the aberrated Airy pattern. It is thus found that tolerances for aberrations are conveniently expressed in terms of $W(x, y)$ when the aberration correction of the system is such that the point image is a good approximation to the Airy pattern; it is therefore useful as a measure of aberration for high quality systems.

It may be mentioned also that the wavefront aberration is obtained directly from several interferometric methods of testing lenses, such as the Twyman and Green interferometer; this provides a useful comparison with the results of computation when the accurate construction of an optical system is in doubt or when it is desired to carry out final correction of spacings or of the figure of optical surfaces.

When the conditions of use do not require such high quality aberration correction, e.g. when the resolution limit of the detector is much greater than the Airy pattern, other measures of aberration quality may be more appropriate. The transverse ray aberrations give a reasonably good impression of a heavily aberrated point image and in later sections of this chapter we show some different ways in which the aberrations can be displayed for this purpose. Also, since the transverse ray aberrations are obtained directly from raytracing they are often used in the detailed stages of computer optimization of lens designs. Measures of image quality based on the diffraction image rather than geometrical optics are described briefly in Chapter 13.

Clearly both methods of describing aberrations have important advantages and it is therefore essential to be familiar with both. As explained in Chapter 2, the wavefronts with which we are dealing are strictly a geometrical optics concept; nevertheless the wavefront aberrations yield information which can be applied to the physical optics problems of diffraction image formation and thus $W(x, y)$ is the essential link between the two domains.

We may note finally how the eikonal as defined in Section 7.2 can be related to the wavefront aberration. When the eikonal is used for basic theoretical studies the origins O and O′ (Fig. 7.2) from which perpendiculars are drawn to the ray segments are usually on the optical axis and are either Gaussian conjugates or, say, the principal foci. However, this need not be so and we could, for example, take them to be the intersections of a finite principal ray with the Gaussian object and image planes, P_0 and P_0' (Fig. 7.3). Then it can be shown (see Chapter 9) that if Q and Q′ are the feet of the perpendiculars from P_0 and P_0' to another finite ray of the pencil the quantity $[P_0P_0'] - [QQ']$ is the wavefront aberration referred to a reference sphere of infinite radius of curvature centred on P_0'. Since $[QQ']$ is the eikonal for origins P_0 and P_0' we have related the two concepts directly.

7.5 Other methods of computing the wavefront aberration

Finite raytracing is the basic method for calculating the exact behaviour of an optical system; from this we can obtain directly the transverse ray aberrations and then, by numerical integration according to eqn (7.16), the wavefront aberration. All other techniques for computing the wavefront aberration or, what is more or less equivalent, one of the characteristic functions defined in Section 7.2, also depend essentially on finite raytracing. But there are several variations, of which one or other may be more convenient, depending on the aim of the computation and on the speed and capacity of the computer available.

One method suggests itself immediately from the definition of wavefront aberration given in Section 7.2. Referring to Fig. 7.3, W at the point $Q_0(x, y, z)$ on the reference sphere is given as the difference between the optical paths from the object point P to Q_0 and from P to O, the centre of the pupil, as in eqn (7.9). We can then trace finite rays from the object point through the system, find their intersections with the reference sphere, sum the segments of optical path length along the rays and use eqn (7.9). To do this we have to make a transfer to a decentred spherical surface, the reference sphere, and we have to obtain the optical path lengths of the ray segments.

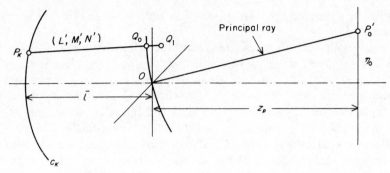

FIG. 7.5. Calculating the intersection of a ray with the final reference sphere

Figure 7.5 illustrates the transfer to the reference sphere; the ray with direction cosines (L', M', N') leaves the final, kth, surface at P_k, of which the coordinates are (x_k, y_k, z_k) in the system with origin at this surface. Let this ray meet the reference sphere at Q_0 and the pupil plane at Q_1 and let the pupil plane be at a distance $\bar{l}$ from the last surface. If the coordinates of Q_1 are $(x_1, y_1, 0)$ and those of Q_0 are (x, y, z) we have

$$
\left.
\begin{aligned}
x_1 &= x_k + \frac{L'}{N'}\,(\bar{l} - z_k) \\[2mm]
y_1 &= y_k + \frac{M'}{N'}\,(\bar{l} - z_k)
\end{aligned}
\right\}
\tag{7.20}
$$

and

$$
\left.
\begin{aligned}
x &= x_1 + L'\Delta_p \\
y &= y_1 + M'\Delta_p \\
z &= N'\Delta_p
\end{aligned}
\right\},
\tag{7.21}
$$

where $\Delta_p = Q_1 Q_0$. To obtain Δ_p we eliminate x, y and z between eqns (7.21) and the equation of the reference sphere, eqn (7.11). The procedure including the choice of sign of a square root is just as in the ordinary finite raytrace of Chapter 4 and we can omit the details; the result is

$$
\Delta_p = \frac{x_1^2 + y_1^2 - 2\eta_0 y_1}{N'z_p - L'x_1 - M'(y_1 - \eta_0) + \{(N'z_p - L'x_1 - M'(y_1 - \eta_0))^2 - x_1^2 - y_1^2 + 2\eta_0 y_1\}^{\frac{1}{2}}},
\tag{7.22}
$$

which completes the transfer to the reference sphere.

Let D_j be the length of the ray segment between surfaces $j - 1$ and j; then it is easily seen that:

$$D_j = \frac{d'_{j-1} - z_{j-1}}{N_j} + \Delta_j, \quad (j = 1, 2, \ldots) \tag{7.23}$$

where the symbols refer to the usual raytracing quantities of Chapter 4 but with the subscript j to show explicitly the surface in question. In the same way the length of the ray segment from the last (kth) surface to the reference sphere is

$$D_p = \frac{l - z_k}{N'_k} + \Delta_p \tag{7.24}$$

and D_1, the segment from the object point to the first surface, is obtained by an obvious modification of eqn (7.23). Then from eqn (7.9) we have

$$W(x, y) = \sum n(\bar{D} - D) \tag{7.25}$$

where $\bar{D}$ refers to the principal ray and D to the ray meeting the reference sphere in the point Q_0 with coordinates (x, y, z).

This result is in a form which shows directly the dependence of the wavefront aberration on the parameters of the ray paths. However, as it stands it is of limited practical use because each of the terms of the right-hand side is of the order of magnitude of the overall length of the optical system, perhaps several hundred millimetres, whereas the difference between them is to be calculated to a fraction of a wavelength; this may make a heavy demand on the computer when rounding-off errors are allowed for, since the computer word length may correspond to only about six decimal digits. We can put eqn (7.25) into a form requiring less numerical precision in the following way. Equation (7.23) can be written as

$$D_j = \frac{d'_{j-1} - z_{j-1} + z_j}{N_j}. \tag{7.26}$$

Then we have

$$W(x, y) = \sum_j n_j(\bar{D}_j - D_j)$$
$$= \sum_j n_j \left\{ \frac{(d'_{j-1} - \bar{z}_{j-1} + \bar{z}_j)}{\bar{N}_j} - \frac{(d'_{j-1} - z_{j-1} + z_j)}{N_j} \right\}. \tag{7.27}$$

Now, making use of the properties of direction cosines we have, if $\bar{N}$ and

N are any two z-direction cosines,

$$
\begin{aligned}
\frac{1}{\overline{N}} - \frac{1}{N} &= \frac{N - \overline{N}}{N\overline{N}} \\
&= \frac{N^2 - \overline{N}^2}{N\overline{N}(N + \overline{N})} \\
&= \frac{\overline{L}^2 + \overline{M}^2 - L^2 - M^2}{N\overline{N}(N + \overline{N})}
\end{aligned} \tag{7.28}
$$

and since L and M are usually much smaller than N this eliminates a loss of precision. Next, substituting this expression in eqn (7.27) we have

$$
W(x, y) = \sum_j n_j \left\{ d'_{j-1} \frac{(\overline{L}^2 + \overline{M}^2 - L^2 - M^2)}{N\overline{N}(N + \overline{N})} + \frac{\overline{z}_j - \overline{z}_{j-1}}{\overline{N}_j} - \frac{(z_j - z_{j-1})}{N_j} \right\}. \tag{7.29}
$$

In this expression we no longer have very large numbers subtracted from each other to give a very small result; all the terms are of the order of magnitude of z, which is typically only a few millimetres, and so at the cost of a slight increase in complexity in the expression the numerical precision is improved to the point at which there is enough precision even for a six decimal digit computer for an optical system of moderate size.

For many methods of image evaluation it is desirable to have either the transverse ray aberration components or the wavefront aberration in the form of a polynomial or a spline in x and y, the pupil coordinates, rather than as sets of values at isolated points. We shall not go into the details of this polynomial fitting process here; the technique of fitting, usually by least squares, is described in texts on numerical methods and subroutines are available on all computers. We assume that it is possible to fit polynomials to the transverse ray aberrations and then to carry out the integration according to eqn (7.16) algebraically on these polynomials, i.e. to evaluate the indefinite integral. An interesting point arises here concerning the integration; the two transverse ray aberrations are in general functions of both variables x and y and they are connected by the property that they are proportional to the partial derivatives of the wavefront aberration (see eqn (7.15)). Thus the integrand in eqn (7.16) must be a total differential, i.e. we must have

$$
\frac{\partial}{\partial y} \delta \xi = \frac{\partial}{\partial x} \delta \eta. \tag{7.30}
$$

However, although this condition must be fulfilled for the actual functions $\delta \xi$ and $\delta \eta$ it does not follow that it will hold for fitted polynomials with terms of any form, since these are approximations; if eqn (7.30) is not fulfilled exactly it will not be possible to calculate $W(x, y)$ as the indefinite integral of $\delta\xi\ dx +$ $\delta\eta\ dy$. Physically this means that different values of W would be obtained according to the choice of the path of integration between the two points A and B in the pupil in eqn (7.16); this is an absurdity which could arise because of the choice of inappropriate terms in the transverse ray aberration polynomials and it can be avoided by constraining these in advance to be of the correct forms. For example, we know from symmetry that for an off-axis pencil W is even in x (see Section 7.6) so that we have

$$W = ax^2 + by^2 + cx^2y + dy^3 + \cdots; \tag{7.31}$$

it follows that $\delta\xi$ and $\delta\eta$ must be of the forms

$$\left.\begin{array}{l} \delta\xi = 2ax + 2cxy + \cdots \\ \delta\eta = 2by + cx^2 + 3dy^2 + \cdots \end{array}\right\} \tag{7.32}$$

apart from a constant factor; no other terms up to the third order are allowed and the coefficients must be related as in eqn (7.32).

Finally we note that for some purposes it may be useful to compute aberration polynomials as functions of three variables, the two pupil coordinates and a third coordinate which is a measure of field angle or image height, rather than having separate polynomials for each field angle.

7.6 The theory of aberration types

So far in this chapter we have discussed aberrations by taking finite ray-tracing as a starting point; thus we have seen how to compute the total aberration through a system as a function of pupil position and field angle, but we have not considered what kinds of aberration are likely to arise in a system and we thus have no indication of the kinds of terms to be expected in an aberration polynomial.

Sir William Hamilton developed a very elegant theory of aberration types, (see the reference in Section 7.1), starting from his definition of the characteristic function and using the rotational symmetry of the optical system to show that only certain terms could possibly occur. These terms correspond to the spherical aberration, coma, etc. which are computed and measured for optical systems. We can develop the theory by observing that the eikonal E of Section 7.2 is nominally a function of four variables, L, M, L', M', these being the x- and y-direction cosines of the object- and image-space segments of a ray. However, the optical system is rotationally symmetric about the z-axis, so that a rigid rotation of the ray about this axis cannot change the value of E; alternatively, E must be independent of a rigid rotation of the x- and

y-axes about the z-axis, so that it must be possible to write E as a function of variables which do not depend on the azimuth of the x- and y-axes. Such variables are the inclination of the ray segments in object and image space to the z-axis, and the angle between these segments, and it is easily seen that these *three* variables are both necessary and sufficient to fix the ray configuration completely, apart from the rigid rotation. We can, of course, choose any functions of these three angles as variables; Hamilton used quantities similar to

$$L^2 + M^2, \quad LL' + MM', \quad L'^2 + M'^2, \tag{7.33}$$

and the first and third of these can be seen to be the squares of the sines of the angles U and U' of the ray segments with the axis; the second is $\cos \theta - \cos U \cos U'$, where θ is the angle between the segments. Hamilton wrote

$$E \equiv E(L^2 + M^2, LL' + MM', L'^2 + M'^2) \tag{7.34}$$

and he then postulated that E could be expanded as a power series in the three variables. It turned out that the first order terms corresponded to Gaussian optics and the higher order terms to aberrations of different kinds.

For our purposes it is useful to carry out a similar expansion in terms of the wavefront aberration W. We start by regarding W as a function of four variables, x, y, the exit (or entrance) pupil coordinates and ξ, η, the coordinates of the object point. Note that for the moment we do not restrict the object point to lie in the plane containing the y- and z-axes. Now on account of the rotation symmetry W must not change if there is a rigid rotation of the ξ-, η- and x-, y-axes around the z-axis, so that it must really be a function of combinations of x, y, ξ, η which are rotation invariant. Such combinations are

$$x^2 + y^2, \quad x\xi + y\eta, \quad \xi^2 + \eta^2; \tag{7.35}$$

it can be verified by direct substitution according to the equations

$$\begin{aligned} x &= x' \cos \theta - y' \sin \theta, & y &= y' \cos \theta + x' \sin \theta \\ \xi &= \xi' \cos \theta - \eta' \sin \theta, & \eta &= \eta' \cos \theta + \xi' \sin \theta \end{aligned} \right\} \tag{7.36}$$

that these combinations are invariant under rotations, and all other rotation invariant combinations must be functions of these since we could not get more than three independent invariants from four variables. Thus W must be a function of the three quantities in eqn (7.35). We now make further use of the symmetry by recalling that, as in Section 7.2, we need only consider object points along the η-axis; we therefore put ξ identically equal to zero and write W as a function of $x^2 + y^2$, $y\eta$ and η^2. Finally, following Hamilton, we assume that W can be expanded as a power series in these variables and we write

$$W(x, y, \eta) \equiv W(x^2 + y^2, y\eta, \eta^2)$$
$$= a_1(x^2 + y^2) + a_2 y\eta + a_3 \eta^2 + b_1(x^2 + y^2)^2 + b_2 y\eta(x^2 + y^2)$$
$$+ b_3 y^2\eta^2 + b_4\eta^2(x^2 + y^2) + b_5 y\eta^3 + b_6\eta^4 + \cdots \qquad (7.37)$$
$$+ \text{third and higher order terms.}$$

The constant term is omitted because in the definition of wavefront aberration we have assumed that both the wavefront and the reference sphere are chosen to pass through the centre of the pupil (see Fig. 7.3), so that W must be zero at the origin of the x, y coordinate system. This being so we note that there can be no terms which are independent of x and y, so that the co-efficients a_3, b_6, etc. of η^2, η^4 and higher powers must all be zero.

Of the remaining terms it can easily be shown that there are $\frac{1}{2}n(n + 3)$ of degree n in the three variables, excluding terms in η^2 alone, as noted above. The two first degree terms (in $x^2 + y^2$ and $y\eta$) have a special significance: it is customary in the Hamiltonian theory to postulate that in the wavefront aberration expansion the reference sphere for each field angle has its centre at the Gaussian image point, i.e. the aberrations are measured taking the Gaussian image point as the ideal image point; now if we refer to eqn (7.18) we see that a longitudinal shift of the centre of the reference sphere contributes a term in z, i.e. in $x^2 + y^2$ to the aberration, so that the presence of such a term indicates that the reference sphere is not correctly centred at the Gaussian image plane. Similarly a non-zero $y\eta$ term would mean a transverse shift of the centre of the reference sphere. Thus the two linear terms do not represent proper aberrations and they are not normally included in the aberration expansion.

The five second degree terms with coefficients b_1 to b_5 inclusive are thus the first group of aberration terms. Their significance will be examined in detail in the next section. The various groups of aberrations have been given several different names; for example, it follows from eqn (7.15) that the b_1 to b_5 group will be of the third degree in the variables x, y and η if they are ex-pressed as transverse ray aberrations, so they are sometimes called the third order aberrations, and succeeding groups are fifth, seventh, etc. orders. Alternatively the b_1 to b_5 group may be called the primary aberrations and succeeding groups secondary, tertiary, etc. Explicit formulae for computing the third order aberrations were first given systematically by L. Seidel in 1856 so they are also called the Seidel aberrations. Seidel's full name was Philip Ludwig von Seidel but he published his work on aberrations as L. Seidel, so it seems best to continue to refer to him thus; some references give L. von Seidel and others simply von Seidel. According to the *Dictionary of Science Biography* (Putnam 1975) his father did not boast a "von" so perhaps Seidel assumed it at some time rather as Englishmen assumed an "Esquire" for greater personal dignity.

In eqn (7.37) x and y are understood to be the coordinates in the pupil axes of the point P_0 in which the ray meets the reference sphere (Fig. 7.3); with this proviso we have an unambiguous definition of all the groups of higher order aberrations according to eqn (7.37). However, it should be pointed out that this is not the only possible way in which an aberration expansion could be written down: we might, for example, take the coordinates in the pupil system as those of the point at which the ray crosses the pupil plane and then a given system would be found to have different coefficients for the higher order aberration terms. The reason for this is that a change such as that proposed would produce a change of order of magnitude α^3 in x or y, where α has the significance of a variable which is linear in x, y and η, and this substituted in any of the Seidel terms produces terms of the 6th degree, i.e. secondary or 5th order aberration changes. By the same argument it can be seen that the Seidel terms are *not* affected by details of the exact significance of the coordinates. Thus the Seidel terms are unique and all of the many systems for computing them lead to essentially the same quantities, apart from constant numerical factors, but there is no such unanimity about higher order aberration terms. However, this is not a serious matter, (a) because all optical designs are finally checked by raytracing and (b) because it is not the custom with present day design techniques to compute many higher order aberrations explicitly as power series terms; the *concept* of higher order aberrations is nevertheless very useful in discussing the possibilities and limitations of an optical design.

It is relatively simple to compute numerically the values of the Seidel terms for a given system; formulae for doing this will be derived in the next chapter. These formulae, and likewise the discussion in the next section of the significance of the aberration terms, will be developed in terms of both ray aberrations and wavefront aberrations since both of these are useful in practice. Similarly we shall use polar coordinates in the pupil instead of x and y when this simplifies the discussion. It is desirable to stress that these alternative descriptions all relate to the same physical things and none is more fundamental than any other. The Seidel aberration formulae in their various forms are sufficiently simple in structure to permit the development of many general theorems about the aberrations which can occur in optical systems and these form a basis for the first approach to a corrected system. This is another reason for the extended study of the Seidel aberrations, but on the other hand all higher order formulae so far developed are relatively complicated and very few general theorems have been found.

7.7 The Seidel aberrations

In this section we shall examine in detail the forms of the five terms of eqn (7.37) in the 4th degree in x, y and η which were called the Seidel aberrations. We shall suppose that the numerical values of the coefficients b_1 to b_5 have been found (by formulae to be given in Chapter 8) and we wish to find the wavefront shapes and ray intersection patterns which result.

The first term, $b_1(x^2 + y^2)^2$ is called *spherical aberration*.[†] The field variable η does not appear in this term so that its effect is constant over the field of the system.[‡] In discussing this and the other aberrations it is sometimes convenient to use ρ, ϕ, polar coordinates in the pupil, as alternatives to x, y, the cartesian coordinates of Fig. 7.3. We have

$$\rho^2 = x^2 + y^2, \qquad \tan \phi = x/y,$$
$$x = \rho \sin \phi, \qquad y = \rho \cos \phi, \qquad (7.38)$$

and the basic variables of the aberration expansion (eqn (7.37)) take the form

$$\rho^2, \quad \rho\eta \cos \phi, \quad \eta^2. \qquad (7.39)$$

Spherical aberration is therefore of the form $b_1\rho^4$ as a wavefront aberration. The form of the deviation of the wavefront from the reference sphere is axially symmetrical about the principal ray; the profile in the y–z section is indicated in Fig. 7.6 and it can be seen that the reference sphere and the wavefront have the same curvature in the centre of the pupil. The y–z section, which contains the principal ray according to our convention in putting $\zeta = 0$ for the Gaussian image point, is called the meridian or tangential section. If we suppose the pencil to be transferred to star space (by a notional aberration-free system), the reference sphere becomes a plane and the meridian section is as in Fig. 7.7. We can also imagine a view of this diagram in the direction

---

[†] This name has no significance in the present context; it came into use for historical reasons which have since been shown to be wrong, but no confusion is caused by keeping to it. The German equivalent *Öffnungsfehler* (aperture aberration) is less inappropriate.

[‡] Such statements as this recur throughout this section but they have meaning only within the context of the aberration expansion theory. If the exact behaviour of a given optical design is determined by raytracing, it is in general found that an aberration depending on $(x^2 + y^2)^2$ is present but that the magnitude of the coefficient varies with field angle; this means that the varying part really ought to be written as another term of the form $\eta^{2n}(x^2 + y^2)^2$, i.e. a higher order aberration. Also it might be found that the coefficient of $(x^2 + y^2)^2$ obtained from raytracing is not equal to the coefficient obtained from the formulae of Chapter 8. This is because the terms obtained by polynomial fitting to the raytrace results are fitted to an arbitrarily chosen set of points in the x, y, η space, whereas the aberration expansion is like a Taylor series with the paraxial region as origin, so that it is fitted to the higher order derivatives at the origin.

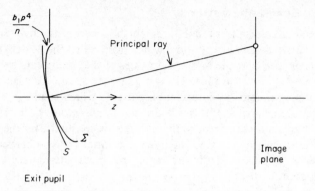

FIG. 7.6. The wavefront shape in primary spherical aberration

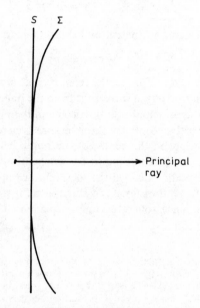

FIG. 7.7. A wavefront in star space with primary spherical aberration

of the arrow, plotting contours or levels of constant wavefront aberration. This would appear as in Fig. 7.8; if each contour corresponds to, say, an increase of one wavelength in wavefront aberration, the radii clearly increase as $N^{\frac{1}{4}}$, where N is the number of wavelengths. Figure 7.8 corresponds to what would be observed in an interferometer for measuring aberrations, such as the Twyman–Green interferometer, the interference fringes corresponding to levels at half wavelength increments.

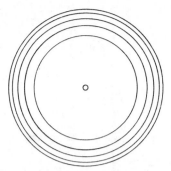

FIG. 7.8. Levels or contours of constant primary spherical aberration

If we now go to the transverse ray aberration representation we find from eqn (7.15) that the aberration components are

$$
\left.
\begin{aligned}
\Delta\xi &= -\frac{4b_1 R}{n} x(x^2 + y^2) \\
\Delta\eta &= -\frac{4b_1 R}{n} y(x^2 + y^2)
\end{aligned}
\right\} .
\tag{7.40}
$$

When, as in the present case, the aberration is symmetrical about the principal ray it is sufficient to consider only one of these, say $\Delta\eta$, and to deal only with rays in the meridian section. We then can plot $\Delta\eta$ as a function of y,

$$
\Delta\eta = -\frac{4b_1 R}{n} y^3
\tag{7.41}
$$

and we obtain Fig. 7.9.

We can also draw the rays in the meridian section as normals to the wavefront and we obtain the very useful representation of Fig. 7.10. The two halves of this figure are not, of course, drawn to the same scale. The rays near the Gaussian image point touch a surface of revolution, the caustic, of which the equation of the meridian section can be found as follows. From eqn (7.41) the equation in η, ζ coordinates of a ray is

$$
\eta = Cy^3 - \frac{(y - Cy^3)}{R}\zeta,
\tag{7.42}
$$

where $C = -4b_1 R/n$. In eqn (7.42) y is a parameter which defines a particular ray of the family of rays in the meridian section. We can find the envelope of this family in the usual way by differentiating eqn (7.42) with respect to y and eliminating y between the result and eqn (7.42). We obtain

$$
\eta = -\frac{2}{3\sqrt{3R}} \cdot \frac{\zeta^{3/2}}{(C(R + \zeta))^{\frac{1}{2}}}
\tag{7.43}
$$

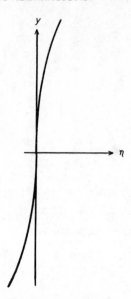

FIG. 7.9. Primary spherical aberration as transverse ray aberration at the
Gaussian image plane

or, if we assume ζ is small compared to R,

$$\eta = -\frac{2}{3\sqrt{3C}}\left(\frac{\zeta}{R}\right)^{3/2}, \tag{7.44}$$

a semi-cubical parabola. Only the upper branch is indicated in Fig. 7.10.†
Since all the rays touch the caustic we might expect a great intensity of light
in this region and this leads us to investigate the distribution of rays in image
planes other than that containing the paraxial focus. It is usual to assume
that the light intensity in an image is obtained according to the geometrical
optics approximation by tracing rays from the object point through a uniform
lattice of points in the entrance pupil and taking the light intensity to be proportional to the density of ray intersections in the chosen image plane. This
interpretation is obviously invalid on the evolute, where the ray intersection
density is infinite, but elsewhere in the image this provides a useful guide to

---

† This caustic surface is, of course, what is called in differential geometry the evolute surface
of the wavefront. The evolute of a surface is in general a surface of two sheets, the loci of
the centres of curvature of the principal sections or, what is the same thing, the envelope of
the normals. The evolute is thus common to all the wavefronts of a pencil, as is obvious
from the way in which it arises in Fig. 7.10. In the case of a surface of revolution, such as a
wavefront with spherical aberration, the second sheet of the evolute degenerates to a single line,
the axis of revolution, which passes through the cusp of the curve defined by eqn (7.43) or (7.44).

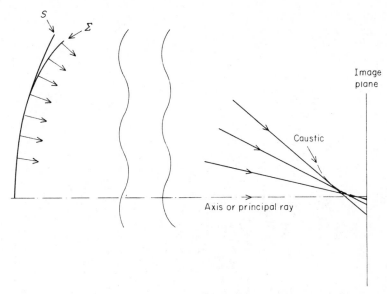

FIG. 7.10. Ray paths in primary spherical aberration; the intersections with the axis (principal ray) give the longitudinal ray aberration

image quality. We can sidestep the difficulty of the infinities by simply plotting isolated points for the individual ray intersections, as in Fig. 7.11; the result is called a *spot diagram* and it gives a very useful impression of the geometrical image quality. The points can be drawn as squares, circles, etc., to indicate the regions of the pupil from which the rays originate. Such diagrams can easily be obtained from a computer graph-plotter programmed to operate directly from the output of a raytracing programme.

The ray intersection density can be calculated by noting that a rectangle $dx.dy$ on the wavefront is delineated by the rays through it as a parallelogram on the image plane and the ray density is inversely proportional to the area of this parallelogram. The four vertices of the parallelogram have the

FIG. 7.11. A spot diagram for spherical aberration

coordinates

$$-\frac{R}{n}\left(\frac{\partial W}{\partial x}, \frac{\partial W}{\partial y}\right)$$

$$-\frac{R}{n}\left(\frac{\partial W}{\partial x} + \frac{\partial^2 W}{\partial x^2}\,\mathrm{d}x, \frac{\partial W}{\partial y} + \frac{\partial^2 W}{\partial x\,\partial y}\,\mathrm{d}x\right)$$

$$-\frac{R}{n}\left(\frac{\partial W}{\partial x} + \frac{\partial^2 W}{\partial x\,\partial y}\,\mathrm{d}y, \frac{\partial W}{\partial y} + \frac{\partial^2 W}{\partial y^2}\,\partial y\right)$$

$$-\frac{R}{n}\left(\frac{\partial W}{\partial x} + \frac{\partial^2 W}{\partial x^2}\,\mathrm{d}x + \frac{\partial^2 W}{\partial x\,\partial y}\,\mathrm{d}y, \frac{\partial W}{\partial y} + \frac{\partial^2 W}{\partial x\,\partial y}\,\mathrm{d}x + \frac{\partial^2 W}{\partial y^2}\,\mathrm{d}y\right)$$

$$\left.\right\} , \quad (7.45)$$

so that the area of the parallelogram is

$$-\frac{R}{n}\begin{vmatrix} \dfrac{\partial^2 W}{\partial x^2} & \dfrac{\partial^2 W}{\partial x\,\partial y} \\[2mm] \dfrac{\partial^2 W}{\partial x\,\partial y} & \dfrac{\partial^2 W}{\partial y^2} \end{vmatrix}\,\mathrm{d}x\,\mathrm{d}y. \qquad (7.46)$$

Thus the ray intersection density is proportional to the reciprocal of the determinant

$$\begin{vmatrix} \dfrac{\partial^2 W}{\partial x^2} & \dfrac{\partial^2 W}{\partial x\,\partial y} \\[2mm] \dfrac{\partial^2 W}{\partial y\,\partial x} & \dfrac{\partial^2 W}{\partial y^2} \end{vmatrix}. \qquad (7.47)$$

This function is plotted for spherical aberration in Fig. 7.12 at several planes of focus near the paraxial focus. We can see from this figure that the point image could hardly be said to be the "best" image at the paraxial focal plane on almost any criterion, i.e. we improve the image by defocusing it. The smallest image will be where the ray from the rim of the pupil meets the caustic, as in Fig. 7.13. Let the pupil radius be a; from eqn (7.42) the rim ray meets the image plane at $\eta = -Ca^3 + a\zeta/R$ to a sufficient approximation, so that from eqn (7.44) the required defocus distance ζ is given by the equation

$$-Ca^3 + \frac{a\zeta}{R} = -\frac{2}{3\sqrt{3C}}\left(\frac{\zeta}{R}\right)^{3/2}. \qquad (7.48)$$

By factorizing or by trial a solution of this cubic in ζ is $3Ca^2R/4$. The point at which the rim rays cross the principal ray is $\zeta = Ca^2R$, so that the smallest image patch is found three-quarters of the way from the paraxial focus to the

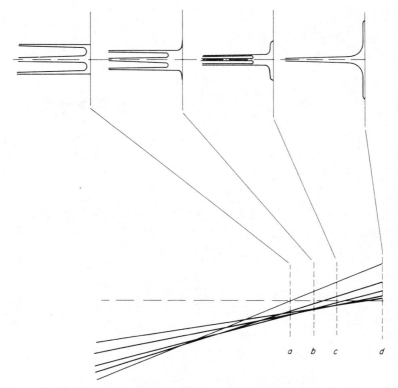

FIG. 7.12. Ray intersection densities for primary spherical aberration at
different planes of focus

rim or marginal focus and its radius is, from eqn (7.44), $Ca^3/4$. This image
patch is called the *disc of least confusion*.

The effect of the focal shift on the size of the image patch can be illustrated
in a slightly different way by noting that it is equivalent to adding a linear
term to eqn (7.41) for the transverse ray aberration:

$$\Delta\eta = -\frac{4b_1 R}{n}y^3 - \frac{\zeta}{R}y. \tag{7.49}$$

Thus Fig. 7.9 has a linear term added and the value of ζ corresponding to
the disc of least confusion is that for which the graph has equal positive and
negative excursions, as in Fig. 7.14. Clearly the addition of a suitably chosen
linear term considerably reduces the size of the image patch, by a factor 4 in
fact.

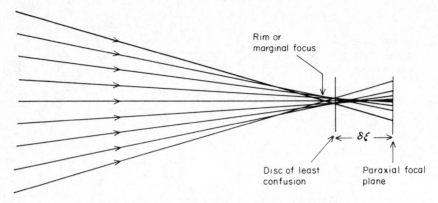

FIG. 7.13. The disc of least confusion for spherical aberration

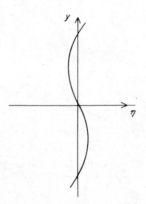

FIG. 7.14. The transverse ray aberration at the disc of least confusion;
this is obtained from Fig. 7.9 by adding a linear term

Returning now to the wavefront aberration representation, we recall from Sections 7.3 and 7.6 that a focal shift corresponds to the addition of a quadratic term to the wavefront aberration; from eqn (7.18) this term is seen to be $(n\zeta/2R^2)(x^2 + y^2)$, where ζ is the focal shift, so that the wavefront aberration with respect to the new reference sphere is

$$W = b_1\rho^4 + \frac{n}{2R^2}\zeta\rho^2, \tag{7.50}$$

and Fig. 7.7 would now appear as in Fig. 7.15. The effect of the shift of focus is to decrease the departure of the wavefront from the reference sphere; this may also be seen from Fig. 7.16, which shows the levels of wavefront aberration. Figures 7.15 and 7.16 are drawn for the focal shift which gives the disc of least

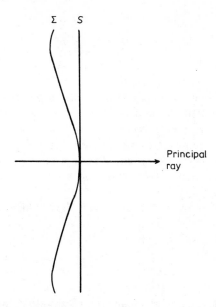

FIG. 7.15. Primary spherical aberration with defocus as a wavefront aberration
in star space; this is obtained from Fig. 7.7 by adding a quadratic term

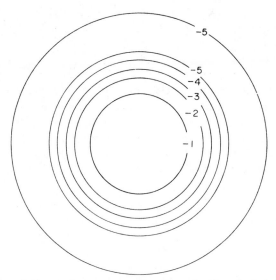

FIG. 7.16. Contours of constant wavefront aberration for spherical aberration
with defocus

confusion, so that the wavefront aberration is

$$W = b_1(\rho^4 - \tfrac{3}{2}a^2\rho^2). \tag{7.51}$$

This amount of defocus completely reverses the sign of the aberration, as can be seen by comparing Figs 7.15 and 7.7. It is not obvious, however, that this is the optimum from a wave-aberration point of view; for example, we might argue that the optimum defocus is that amount which minimizes the mean square wave aberration over the pupil and in fact this is found to be the correct choice according to one of the criteria in Chapter 13. The point that wave and ray aberration criteria do not generally agree is worth making here, although the extent of the disagreement will not be fully apparent until Chapter 13.

The second term in the expansion of eqn (7.37) which we have to consider is *coma*, $b_2 y\eta(x^2 + y^2)$, or, in polar coordinates, $b_2\rho^3\eta \cos \phi$. On account of the factor η this aberration is absent on the axis and it increases linearly with the field angle or distance. If we consider the wavefront shape for a particular point in the field only, the coordinate η can be put together with b_2 as a constant coefficient and the variable part is $y(x^2 + y^2)$. The cross-section of the wavefront in star space across a diameter in azimuth ϕ has a cubic shape and the wavefront therefore appears as in Fig. 7.17. The levels of constant aberration can easily be found to be as in Fig. 7.18.

The transverse ray aberration components are:

$$\left.\begin{aligned}
\Delta\xi &= -\frac{2R}{n}b_2\eta xy \\
\Delta\eta &= -\frac{R}{n}b_2\eta(x^2 + 3y^2)
\end{aligned}\right\}; \tag{7.52}$$

the η component taken across the x and y directions appears as in Fig. 7.19. Here we note that although the wavefront aberration is identically zero across the x-section, or *sagittal* section, the transverse ray aberration does not vanish because the wavefront has a finite gradient in the y direction across the x-axis. It is more usual to write the transverse ray aberrations as functions of polar coordinates in the pupil:

$$\left.\begin{aligned}
\Delta\xi &= -\frac{R}{n}b_2\eta\rho^2 \sin 2\phi \\
\Delta\eta &= -\frac{R}{n}b_2\eta\rho^2(2 + \cos 2\phi)
\end{aligned}\right\}; \tag{7.53}$$

if we consider for the moment only rays from a circle of radius ρ in the pupil it can be seen that eqns (7.53) are the parametric equations of a circle in the

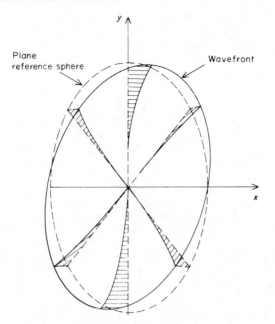

FIG. 7.17. The wavefront shape for primary coma

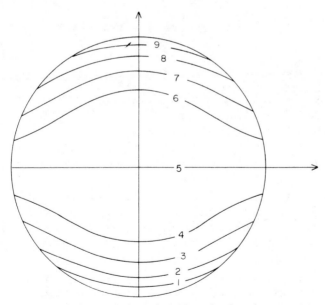

FIG. 7.18. Contours of wavefront aberration for primary coma

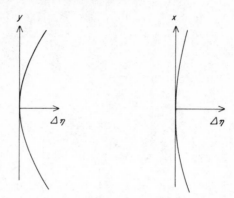

FIG. 7.19. The η-component of transverse ray aberration for primary coma, plotted across the y and x sections of the pupil

image plane, the parameter being ϕ; the centre of the circle is at a distance $2Rb_2\eta\rho^2/n$ from the Gaussian image point O'_1 and its radius is $Rb_2\eta\rho^2/n$ as in Fig. 7.20. Now if ρ is allowed to vary, a series of circles is taced out with

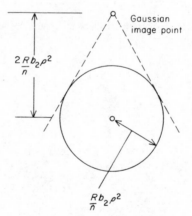

FIG. 7.20. The coma circle

varying radii and centres, each with the tangents from O'_1 enclosing an angle of 60°, so that the whole appears as in Fig. 7.21. Each circle is traced out twice for one traversal of the pupil, on account of the appearance of the double angle in eqns (7.53). This figure does not give a very good impression of the ray density but nonetheless it indicates the reason for the name coma.

Figure 7.22 is a spot diagram for coma and it can be seen that there is a very heavy concentration of light near the Gaussian image point but relatively little light in the largest circles. This also appears from the ray inter-

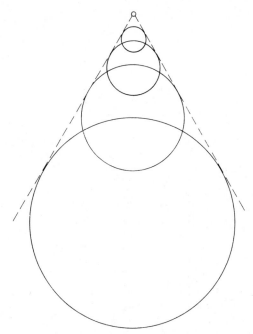

FIG. 7.21. Transverse ray aberration for coma

section density (Fig. 7.23). The striking thing about a comatic image is the marked asymmetry, in general a very undesirable characteristic, particularly where position measurements have to be made, as in astrometry. The scale of the whole coma pattern increases linearly with field angle and, from eqns (7.53), quadratically with the pupil radius.

In view of the improvement in image quality with spherical aberration obtained by a shift of focus it is natural to ask whether the same applies to coma. In terms of wavefront shape a focal shift corresponds to the addition of a quadratic term but since the section of the wavefront in the y–z plane is cubic in shape (Fig. 7.17), i.e. the wavefront aberration is an odd function of y, it is clear that adding an even term (y^2) can only make things worse for negative y-values if it improves them for positive values. The ray picture is more complicated but the same conclusion is usually accepted, that the comatic image is not improved by a shift of focus. On the other hand it can be seen that the ray intersections in the spot diagram are all on one side of the Gaussian image point so that the "best" image centre on any criterion (centre of gravity of the ray intersections, etc.) is shifted laterally, in the η direction. This corresponds to the addition of a linear term in y to the wavefront aberration as in eqn (7.18), so that the wavefront y-section in star space would appear

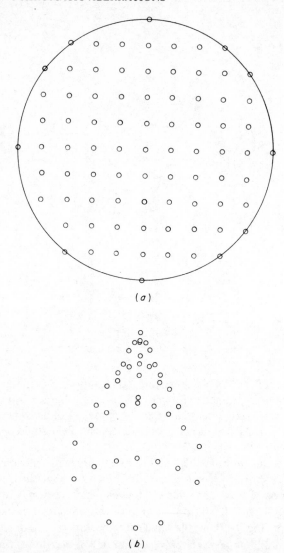

FIG. 7.22. Spot diagram for primary coma; (a) pupil, (b) image plane

as in Fig. 7.24 and the contours of wavefront aberration as in Fig. 7.25; the effect of a suitable magnitude of linear term is clearly to decrease the general departure of the wavefront from the reference sphere. This corresponds to what is usually called the addition of tilt fringes in an interferometric test. Of course, these lateral shifts do not really improve the image in the sense that a focal shift gives an improvement; we are merely selecting a point other than the

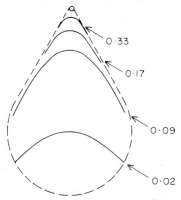

FIG. 7.23. The ray intersection density for coma; the scale of the contours is arbitrary

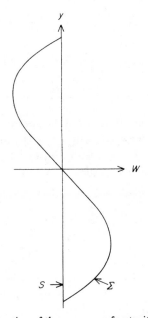

FIG. 7.24. Tangential section of the coma wavefront with linear term added

Gaussian image point to represent the best centre of light concentration in the point image.

The third Seidel aberration, $b_3 y^2 \eta^2$ in eqn (7.37), is called *astigmatism*, again a name which is not particularly appropriate. There is no wavefront aberration in the x-section (sagittal section) but in the y-section (meridian section)

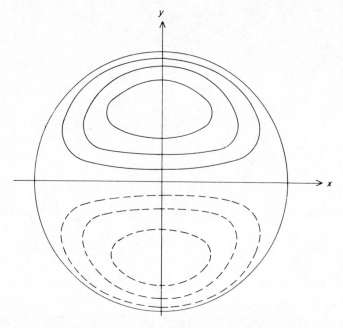

Fig. 7.25. Contours of coma wavefront with linear term

there is an increment of curvature, since the aberration depends quadratically on y. Figure 7.26 shows the wavefront shape in star space in the representation we have already used. The increment in curvature in the y-section depends on the square of the field angle. The components of ray aberration are, from eqn (7.15),

$$\left.\begin{array}{l} \Delta\xi = 0 \\[2mm] \Delta\eta = -\dfrac{2R}{n}b_3\eta^2 y \end{array}\right\}, \tag{7.54}$$

so that all rays pass through a line in the y-section of the image plane of length $4Rb_3\eta^2 a/n$, where a is the pupil radius. Since astigmatism is an even function of the pupil variables we try the effect of a focal shift on the image; as for spherical aberration, this is a term $(n\zeta/2R^2)(x^2 + y^2)$ where ζ is the focal shift, and the ray aberration components become

$$\left.\begin{array}{l} \Delta\xi = -\dfrac{\zeta}{R}x \\[2mm] \Delta\eta = -\left(\dfrac{\zeta}{R} + \dfrac{2R}{n}b_3\eta^2\right)y \end{array}\right\}. \tag{7.55}$$

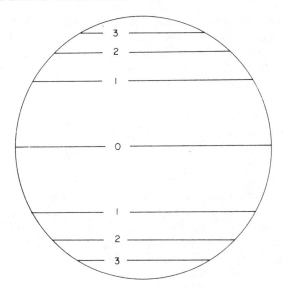

FIG. 7.26. Contours of astigmatism wavefront

If ζ, the amount of focal shift, is chosen so that

$$\frac{\zeta}{R} = -\frac{2R}{n} b_3 \eta^2 \tag{7.56}$$

the coefficient of y in eqn (7.55) vanishes and all the rays pass through a line in the ξ-section of the image plane of length $4Rb_3\eta^2 a/n$. Thus the three-dimensional distribution of rays in the image region is such that they all pass through two lines at right angles, as in Fig. 7.27. The two lines are called the astigmatic foci or astigmatic focal lines. The spacing between them increases as the square of the field angle and so does the length of each line. The focal lines are at the centres of curvature of the wavefront for the x- and y-sections,

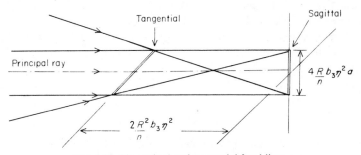

FIG. 7.27. The sagittal and tangential focal lines

i.e. they are portions of the evolute of the wavefront. We have defined the primary aberrations as being of the 4th degree in the pupil and field variables jointly so that, since we have η^2 in the aberration term, we can only have quadratic terms in x and y. Thus there can be no higher degree terms to indicate more of the evolute shape within the limitations of the primary aberration theory.

The best image is usually taken to be halfway between the astigmatic foci; from eqns (7.55) it is easily seen that the rays pass through a circular patch of diameter $2Rb_3\eta^2a/n$, i.e. half the length of a focal line, and this is also sometimes called the disc of least confusion. The spot diagram for this focal position is a pattern of dots uniformly distributed over the circle. The vertical line in Fig. 7.27 can be regarded as the focus for rays from the x-section of the pupil, since $\Delta\xi$ is zero at this focus, and the horizontal line is the focus for the y-section. Thus the vertical line is called the sagittal astigmatic focal line and the horizontal line is called the meridian or tangential focal line; it is necessary to guard against confusion here because the sagittal focal line is actually in the meridian section and vice-versa.

The fourth primary aberration term, $b_4\eta^2(x^2 + y^2)$, has the pupil dependence of a focal shift (ρ^2) but it depends on the square of the field angle; in fact the focus is easily seen to be at a distance $2R^2b_4\eta^2/n$ to the left of the Gaussian image plane, so that the images are true point images on a curved image surface of curvature $4R^2b_4/n$ as indicated in Fig. 7.28. Thus, this aberration is called *field curvature*. Astigmatism and field curvature are often grouped together since they both depend on the square of the field angle. If both b_3 and b_4 coefficients are non-zero then the sagittal focal lines are formed on a curved image surface instead of on the Gaussian image plane and the meridian focal lines are formed on another surface of greater or less curvature, depending on the signs and magnitudes of b_3 and b_4. Figure 7.29 indicates the effect in the η-section of the image region. The surfaces on which the images are formed are called the sagittal and tangential image surfaces. It can be seen that if an object consists of circles concentric with the axis and radial lines, the circles will appear sharp on the tangential focal surface and the radial lines will appear sharp on the sagittal focal surface, as in Fig. 7.30.

The last aberration term in eqn (7.37) is $b_5y\eta^3$ and from eqn (7.18) it follows that this has the effect of a shift of the centre of the reference sphere in the η-direction, i.e. the image point is displaced in the direction of the axis by an amount

$$\Delta\eta = -\frac{R}{n}b_5\eta^3. \qquad (7.57)$$

This is not simply a change in magnification since the shift is proportional to η^3. The effect is that the shape of the image is distorted and consequently this

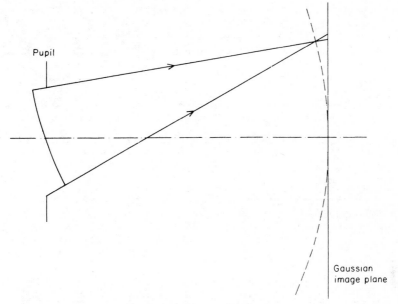

FIG. 7.28. Curvature of the field

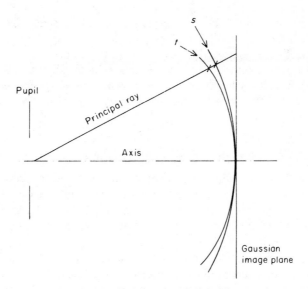

FIG. 7.29. Astigmatic curved fields

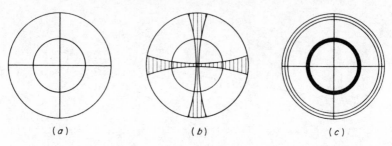

FIG. 7.30. Images of a spoked wheel; (a) the object, (b) image on the tangential focal surface, (c) image on the sagittal focal surface

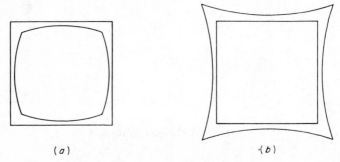

FIG. 7.31. Distortion; (a) barrel, (b) pincushion

aberration is known as *distortion*. We can see the general effect by taking a figure such as a square placed symmetrically with respect to the optical axis and shifting each point radially by an amount proportional to the cube of its distance from the centre as in Fig. 7.31; the original square is shown together with the image, (a) when b_5 is positive, in which case the effect is called barrel distortion, and (b) when b_5 is negative, pincushion distortion.

7.8 Mixed and higher order aberrations

In Section 7.7 we considered the effects of each of the Seidel aberrations separately in order to see how they appear in the image; in practice, of course, they occur in combination with each other and with higher order aberrations. The proportions are determined by the relative magnitudes and signs of the aberration coefficients, $b_1, b_2, \ldots, c_1, c_2, \ldots$. The choice of sign for the coefficients depends on our convention, implied by our definition in Section 7.2 of wavefront aberration as the optical path nQ_0Q (Fig. 7.3), that wavefront aberration is positive when the wavefront is in advance of the reference sphere for positive y in the pupil. This is merely a convention and not all authors adopt it; it has the advantage that the wavefront aberrations come out positive for a biconvex lens with conjugates and pupils in the more usual positions. In agreement with tradition a system with such positive aberrations is said to be *undercorrected*, whereas if one or other of the aberrations is negative the system is said to be *overcorrected* for this aberration.

The analysis into aberration terms appears naturally when the values of the primary aberration coefficients are computed numerically for a given system, as described in Chapter 8. On the other hand, if the properties of the system are determined by raytracing the results appear as the combined effects of all aberrations, including higher order terms; they can still be presented in any of the ways described in Section 7.7, e.g. transverse ray aberration graphs across the meridian and sagittal sections, wavefront aberration graphs for the same sections, levels of wavefront aberration, spot diagrams, etc., but they are not necessarily analysed in detail into simple aberration terms. It is frequently useful to use a broader analysis into general types; thus ignoring for the moment variations with field angle, if the aberration has complete symmetry about the centre of the pupil it would be classed as spherical aberration-like, if the wavefront or the ray congruence has two planes of symmetry, the meridian and sagittal planes, it is astigmatic, and if there is only one plane of symmetry, the meridian plane, the aberration is coma-like. The components of these three degrees of symmetry can usually easily be seen on a graphical representation and they can also be roughly sorted out as follows. Suppose that the wavefront aberration for a given field angle is $W(x, y)$ where the y-axis is, as usual, in the meridian section; here the function $W(x, y)$ may exist only as isolated points computed as in Section 7.5, it may be in graphical form or it may be a polynomial or other fitted function. Then clearly the coma-like component of aberration is $\frac{1}{2}\{W(x, y) - W(x, -y)\}$ and the remainder is $\frac{1}{2}\{W(x, y) + W(x, -y)\}$, which is, say, $W_1(x, y)$. Then $W_1(x, y) - W_1(\sqrt{x^2 + y^2}, 0)$ is the astigmatism-like component and $W_1(\sqrt{x^2 + y^2}, 0)$ is the spherical aberration. It should be emphasized that this decomposition is not rigorous; for example, the "spherical aberration" so found includes not only all higher order spherical aberration terms but also simple defocus or field curvature. However, some such procedure carried out on either ray or wavefront aberration is often adequate to characterize the state of correction of a design for many purposes.

8 Calculation of the Seidel Aberrations

8.1 Addition of aberration contributions

We saw in Section 7.5 that the total wavefront aberration could be written as a sum of contributions from different parts of the system. Consider any oblique pencil and let the principal ray in any chosen space meet the refracting surface at $\bar{P}$ and the axis at $\bar{A}$, as in Fig. 8.1; let O be the paraxial image point in this space and let a reference sphere be drawn with radius $O\bar{A}$; let any other ray of the pencil meet the refracting surface at Q and the reference sphere at A. Then the total wavefront aberration between this ray and the reference ray can be written

$$W = \sum \{n'(\bar{P}\bar{A}' - QA') - n(\bar{P}\bar{A} - QA)\}, \tag{8.1}$$

the summation being taken over all the refracting surfaces;† this expression can easily be seen to be equivalent to eqn (7.25) but we have now put it in a form which shows how an increment of aberration arises at each refraction; thus, if a particular term were zero we should be able to say that the corresponding surface made no contribution to the aberration for the rays in question. The terms in the sum in eqn (8.1) cannot, however, be regarded as arising simply at the respective refracting surfaces because it can be seen that in general a change in the radius of the reference sphere will slightly redistribute the contributions from neighbouring terms, although there will be no net change in the sum. Thus these terms may be regarded as partly due to the refracting surfaces and partly due to the air or glass spaces.

† The primed points are omitted from the figure for clarity.

When we wish to calculate the Seidel aberration terms, i.e. the expressions with coefficients b_1 to b_5 in eqn (7.37), we are naturally led to consider whether they can be expressed similarly as sums through the system. It turns out to be possible to do this by expanding eqn (8.1) in powers of the variables used in Section 7.6 and picking out those terms of the right degree in these variables. The result, leaving aside for the moment the question of convergence of the

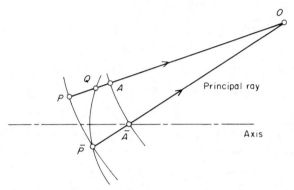

FIG. 8.1. Optical paths difference introduced at a refracting surface

expansion, is a summation over the system and the terms are found to be independent of the radii of the reference spheres, i.e. varying the radius of a reference sphere affects only higher order terms than the Seidel terms. The terms in the Seidel sums are then associated unambiguously with individual surfaces and we speak of surface-by-surface summation of aberrations. The magnitudes of the aberrations will depend on the constructional parameters (radii, refractive indices and spacings) and on the conjugates (object and aperture stop positions).

8.2 Derivation of the Seidel aberration formulae

In order to keep track of the different orders of magnitude of quantities occurring in the derivation we define a quantity μ which shall have mathematically the order of magnitude of x, y and η, the basic quantities of the aberration expansion, eqn (7.37); this is a mathematically defined quantity only, so we do not specify its dimensions in the physical sense. We can then say that the paraxial terms in eqn (7.37) are $O(\mu^2)$ and the Seidel aberration terms are $O(\mu^4)$ whete the $O(\ )$ notation is used in its usual sense.

We first replace eqn (8.1) by a similar one which will enable us to express the Seidel aberrations in terms of surface contributions. Referring again to Fig. 8.1, let $\bar{P}P$ be another reference sphere, also with centre at O but with radius $O\overline{P}$ (and similarly for the primed space), and let s be the difference between the radii of the two reference spheres; thus $s = O\overline{P} - O\overline{A}$; the principal ray $\overline{P}\overline{A}$ must pass close to O, the paraxial image point, and the perpendicular distance from O to this ray is a transverse ray aberration and is therefore of order μ^3 (or perhaps of even higher order of magnitude). Thus the angle between $O\overline{P}$ and the ray $\overline{P}\overline{A}$ is $O(\mu^3)$ and we have

$$\overline{P}\overline{A} - s = O(\mu^6). \tag{8.2}$$

Similarly we can show that

$$PA - s = O(\mu^6) \tag{8.3}$$

and thus for the second term in the summation in eqn (8.1) we have

$$\begin{aligned} \overline{P}\overline{A} - QA &= \overline{P}\overline{A} - PA + PQ \\ &= PQ + O(\mu^6) \end{aligned} \tag{8.4}$$

on using eqns (8.2) and (8.3); similarly for the primed symbols. Substituting this result in eqn (8.1) we have

$$W = \sum \{n'P'Q - nPQ\} + O(\mu^6) \tag{8.5}$$

and by definition terms of order μ^6 do not belong with the primary or Seidel aberrations. We have now expressed the total primary aberration as a surface-by-surface sum, since each term in the summation in eqn (8.5) belongs clearly and unambiguously to the particular surface on which the point Q lies. In other words, we have shown that the choice of the radius of the reference sphere only affects higher order aberrations. This result is some-times called the summation theorem for primary aberrations.

We now have to develop this sum of surface contributions in terms of the construction of the system and other relevant parameters; in what follows we neglect terms of order higher than μ^4 without repeating the above justifica-tion.† In order to carry out this development we draw the radii $\overline{P}O$ and P_1O of the reference sphere, where OP_1 passes through Q as in Fig. 8.2; we can write eqn (8.5) in the form

$$W = \sum \Delta n(PQ) + O(\mu^6), \tag{8.6}$$

where Δ signifies the increment on refraction of the quantity which follows Δ and so

$$W = \sum \Delta n(R + PQ - R) + O(\mu^6) \tag{8.7}$$

---

† At this point it may be worthwhile to emphasize that the higher order aberrations are usually not *numerically* negligible in real optical systems, although we are going to neglect them in deriving the Seidel formulae. However, if we reduce the aperture and the field angle sufficiently we can always make the higher order terms as small as we please compared to the Seidel terms, just as by the same process the Seidel terms can be made negligible com-pared to the Gaussian quantities.

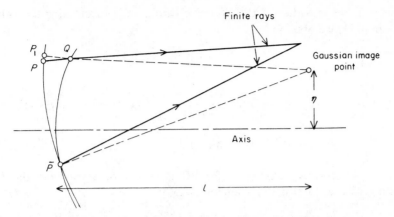

FIG. 8.2. Optical path difference to the approximation of the primary aberrations

where $R = \bar{P}O = P_1O$; but we have

$$PQ = P_1Q + O(\mu^8), \qquad (8.8)$$

so that

$$W = \sum \Delta n(\bar{P}O - QO) + O(\mu^6). \qquad (8.9)$$

Equation (8.9) is to be used as the basis of the aberration expansion. We set up our usual coordinate system with origin at the pole of the refracting surface; let (x, y_1, z) be the coordinates of Q, let $(0, \bar{y}, \bar{z})$ be $\bar{P}$, let l be the Gaussian conjugate distance and η the Gaussian image height. Then we have

$$\bar{P}O - QO = \{(l - \bar{z})^2 + (\eta - \bar{y})^2\}^{\frac{1}{2}} - \{(l - z)^2 + (\eta - y_1)^2 + x^2\}^{\frac{1}{2}}. \qquad (8.10)$$

If the second square root is expanded by the binomial theorem up to and including terms $O(\mu^4)$ we obtain

$$QO = l - z + \frac{x^2 + y_1^2 - 2\eta y_1 + \eta^2}{2l}$$

$$+ \frac{z}{2l^2}(x^2 + y_1^2 - 2\eta y_1 + \eta^2)$$

$$- \frac{1}{8l^3}(x^2 + y_1^2 - 2\eta y_1 + \eta^2)^2. \qquad (8.11)$$

But we have

$$z = \tfrac{1}{2}c(x^2 + y_1^2) + \tfrac{1}{8}c^3(x^2 + y_1^2)^2 \qquad (8.12)$$

where c is the curvature of the refracting surface, and substituting this value in eqn (8.11) we obtain

$$QO = l - \frac{1}{2}c(x^2 + y_1^2) + \frac{x^2 + y_1^2 - 2\eta y_1 + \eta^2}{2l}$$

$$- \frac{1}{8}c^3(x^2 + y_1^2)^2 + \frac{1}{4}\frac{c}{l^2}(x^2 + y_1^2)(x^2 + y_1^2 - 2\eta y_1 + \eta^2)$$

$$- \frac{1}{8l^3}(x^2 + y_1^2 - 2\eta y_1 + \eta^2)^2. \tag{8.13}$$

The corresponding expression for $\bar{P}O$ is clearly obtained from this by putting x equal to zero and changing y_1 to $\bar{y}$:

$$\bar{P}O = l - \frac{1}{2}c\bar{y}^2 + \frac{\bar{y}^2 - 2\eta\bar{y} + \eta^2}{2l} - \frac{1}{8}c^3\bar{y}^4$$

$$+ \frac{1}{4}\frac{c}{l^2}\bar{y}^2(\bar{y}^2 - 2\eta\bar{y} + \eta^2) - \frac{1}{8l^3}(\bar{y}^2 - 2\eta\bar{y} + \eta^2)^2. \tag{8.14}$$

Substituting from eqns (8.13) and (8.14) into eqn (8.9) we obtain for the quadratic terms in μ,

$$W = \sum \Delta n\left\{\frac{1}{2}\left(c - \frac{1}{l}\right)(x^2 + y_1^2 - \bar{y}^2) + \frac{\eta}{l}(y_1 - \bar{y})\right\} + O(\mu^4) \tag{8.15}$$

$$= \sum \left[\frac{1}{2}(x^2 + y_1^2 - \bar{y}^2)\Delta n\left(c - \frac{1}{l}\right) + \frac{y_1 - \bar{y}}{h}\Delta\frac{nh\eta}{l}\right]. \tag{8.16}$$

where h is the incidence height of the paraxial ray from the axial object point. The first difference term vanishes by eqn (3.37) and the second vanishes since $nh\eta/l$ is the Lagrange invariant. It is, of course, not surprising that the wavefront aberration terms which are linear and quadratic in the pupil variables should vanish identically because we are referring the surface aberration contributions to reference spheres centred on the Guassian image points. We are thus left with the $O(\mu^4)$ terms, which must turn out to be primary aberrations.

In order to obtain the fourth power terms in eqn (8.9) we shall make a simple transformation of some of the expressions in eqns (8.13) and (8.14) before substituting in eqn (8.9). The quantities such as x, y_1, $\bar{y}$ etc. which appear in these fourth power terms can now be replaced by their paraxial equivalents, since this change can only affect the 6th degree terms. We have for the Lagrange invariant

$$H = n(\bar{h}u - h\bar{u}) \tag{8.17}$$

from eqn (6.5), so that on division by h and writing $\bar{y}$ for $\bar{h}$,

$$\frac{\eta}{l} = \bar{y}\left(\frac{1}{l} - \frac{1}{\bar{l}}\right); \tag{8.18}$$

Also we put

$$y = y_1 - \bar{y} \tag{8.19}$$

so that paraxially y corresponds to the quantity h, and we substitute from eqns (8.18) and (8.19) to eliminate η and y_1 in the expression

$$\frac{x^2 + y_1^2 - 2\eta y_1 + \eta^2}{l^2} \tag{8.20}$$

which occurs several times in eqns (8.13) and (8.14); the result after some reduction is

$$\frac{x^2 + y_1^2 - 2\eta y_1 + \eta^2}{l^2} = \frac{x^2 + y^2}{l^2} + \frac{2y\bar{y}}{l\bar{l}} + \frac{\bar{y}^2}{\bar{l}^2}; \tag{8.21}$$

by a similar operation we find

$$x^2 + y_1^2 = x^2 + y^2 + 2y\bar{y} + \bar{y}^2. \tag{8.22}$$

Substituting these expressions in eqn (8.13) we obtain for the 4th degree terms in QO,

$$\left(-\frac{1}{8}c^3 + \frac{1}{4}\frac{c}{l^2} - \frac{1}{8l^3}\right)(x^2 + y^2)^2 + \left(-\frac{1}{2}c^3 + \frac{c}{2l\bar{l}} + \frac{c}{2l^2} - \frac{1}{2l^2\bar{l}}\right)\bar{y}y(x^2 + y^2)$$

$$+\left(-\frac{1}{2}c^3 + \frac{c}{l\bar{l}} - \frac{1}{2l^2\bar{l}}\right)\bar{y}^2 y^2 + \left(-\frac{1}{4}c^3 + \frac{1}{4}\frac{c}{\bar{l}^2} + \frac{1}{4}\frac{c}{l^2} - \frac{1}{4}\frac{1}{l^2\bar{l}}\right)\bar{y}^2(x^2 + y^2)$$

$$+\left(-\frac{1}{2}c^3 + \frac{1}{2}\frac{c}{l\bar{l}} + \frac{1}{2}\frac{c}{\bar{l}^2} - \frac{1}{2}\frac{1}{\bar{l}^3}\right)\bar{y}^3 y + \left(-\frac{1}{8}c^3 + \frac{1}{4}\frac{c}{\bar{l}^2} - \frac{1}{8}\frac{l}{\bar{l}^4}\right)\bar{y}^4; \tag{8.23}$$

similarly for the 4th degree terms in $\bar{P}O$ we obtain, by substituting in eqn (8.14), or, more simply, by putting $x = y = 0$ in eqn (8.23)

$$\left(-\frac{1}{8}c^3 + \frac{1}{4}\frac{c}{\bar{l}^2} - \frac{1}{8}\frac{l}{\bar{l}^4}\right)\bar{y}^4. \tag{8.24}$$

Finally the 4th power terms in $PO - QO$ are obtained by subtracting eqn (8.23) from (8.24); this gives simply the first five terms of eqn (8.23) with reversed sign. Thus reverting back to eqn (8.9) we have for the primary aberration of the complete system,

$$
\begin{aligned}
W = \sum_{\substack{\text{all} \\ \text{surfaces}}} \Bigg[& \Delta n \left(\frac{1}{8} c^3 - \frac{c}{4l^2} + \frac{1}{8l^3} \right) (x^2 + y^2)^2 \\
& + \Delta n \left(\frac{1}{2} c^3 - \frac{c}{2l\bar{l}} - \frac{1}{2} \frac{c}{l^2} + \frac{1}{2l^2\bar{l}} \right) \bar{y} y \, (x^2 + y^2) \\
& + \Delta n \left(\frac{1}{2} c^3 - \frac{c}{l\bar{l}} + \frac{1}{2l\bar{l}^2} \right) \bar{y}^2 y^2 \\
& + \Delta n \left(\frac{1}{4} c^3 - \frac{c}{4\bar{l}^2} - \frac{c}{4l^2} + \frac{1}{4l\bar{l}^2} \right) \bar{y}^2 (x^2 + y^2) \\
& + \Delta n \left(\frac{1}{2} c^3 - \frac{c}{2l\bar{l}} - \frac{c}{2\bar{l}^2} + \frac{1}{2\bar{l}^3} \right) \bar{y}^3 y \Bigg].
\end{aligned}
\tag{8.25}
$$

By eqn (8.18) $\bar{y}$ at any one surface is proportional to the image height η, although with a different proportionality constant at each surface, so that $\bar{y}$ can be regarded as a field variable; then the five terms in eqn (8.25) can be identified with the five Seidel aberrations of Chapter 7, since they have similar dependences on the pupil and field variables. We shall next take each term in turn and rearrange it in a form corresponding to the expressions usually used. We note that we can add to any term a quantity which is invariant on refraction, since the operator Δ will remove it, so we add to the first term the quantity

$$
- \frac{1}{8} \sum_{\substack{\text{all} \\ \text{surfaces}}} \Delta n \left(c - \frac{1}{l} \right) c^2 (x^2 + y^2)^2
\tag{8.26}
$$

giving for this term

$$
\begin{aligned}
\frac{1}{8} \sum \Delta n \left\{ c^3 - \frac{2c}{l^2} + \frac{1}{l^3} - c^3 + \frac{c^2}{l} \right\} (x^2 + y^2)^2 \\
= \frac{1}{8} \sum \Delta \frac{n}{l} \left(c - \frac{1}{l} \right)^2 (x^2 + y^2)^2 \\
= \frac{1}{8} \sum n^2 \left(c - \frac{1}{l} \right)^2 (x^2 + y^2)^2 \Delta \left(\frac{1}{nl} \right).
\end{aligned}
\tag{8.27}
$$

Similarly if we add

$$
- \frac{1}{2} \sum \Delta n c^2 \left(c - \frac{1}{l} \right) \bar{y} y (x^2 + y^2)
\tag{8.28}
$$

to the second term in eqn (8.25) we obtain

$$\frac{1}{2}\sum n^2\left(c - \frac{1}{\bar{l}}\right)\left(c - \frac{1}{l}\right)\bar{y}y(x^2 + y^2)\Delta\left(\frac{1}{nl}\right). \tag{8.29}$$

We add

$$-\frac{1}{2}\sum \Delta nc^2\left(c - \frac{1}{l}\right)\bar{y}^2y^2 \tag{8.30}$$

to the third term to obtain

$$\frac{1}{2}\sum n^2\left(c - \frac{1}{\bar{l}}\right)^2\bar{y}^2y^2\Delta\left(\frac{1}{nl}\right). \tag{8.31}$$

We add

$$-\frac{1}{4}\sum \Delta nc^2\left(c - \frac{1}{l}\right)\bar{y}^2(x^2 + y^2) \tag{8.32}$$

to the fourth term in eqn (8.25), giving

$$\frac{1}{4}\sum \Delta n\left(\frac{c^2}{l} - \frac{c}{l^2} - \frac{c}{\bar{l}^2} + \frac{1}{\bar{l}^2 l}\right)\bar{y}^2(x^2 + y^2)$$

$$= \frac{1}{4}\sum \Delta n\left\{\frac{1}{l}\left(c - \frac{1}{\bar{l}}\right)^2 - c\left(\frac{1}{l} - \frac{1}{\bar{l}}\right)^2\right\}\bar{y}^2(x^2 + y^2)$$

$$= \frac{1}{4}\sum n^2\left(c - \frac{1}{\bar{l}}\right)^2\bar{y}^2(x^2 + y^2)\Delta\left(\frac{1}{nl}\right)$$

$$\qquad\qquad -\frac{1}{4}\sum n^2\left(\frac{1}{l} - \frac{1}{\bar{l}}\right)^2\bar{y}^2(x^2 + y^2)\Delta\left(\frac{c}{n}\right) \tag{8.33}$$

The reduction of the final term is somewhat more laborious. As before we add

$$-\frac{1}{2}\sum \Delta nc^2\left(c - \frac{1}{l}\right)\bar{y}^3y, \tag{8.34}$$

giving for the fifth term in eqn (8.25)

$$\frac{1}{2}\sum \Delta n\left(\frac{c^2}{l} - \frac{c}{l\bar{l}} - \frac{c}{\bar{l}^2} + \frac{1}{\bar{l}^3}\right)\bar{y}^3y$$

$$= \frac{1}{2}\sum n\left(c - \frac{1}{\bar{l}}\right)\Delta\left(\frac{c}{l} - \frac{1}{\bar{l}^2}\right)\bar{y}^3y$$

$$= \frac{1}{2}\sum \frac{n(c - 1/\bar{l})}{n(c - 1/l)}\Delta n\left(-\frac{c}{\bar{l}^2} - \frac{c}{l^2} + \frac{c^2}{l} + \frac{1}{l\bar{l}^2}\right)\bar{y}^3y$$

$$= \frac{1}{2}\sum \frac{n^3(c - 1/\bar{l})^3}{n(c - 1/l)}\bar{y}^3y\Delta\left(\frac{1}{nl}\right) - \frac{1}{2}\sum \frac{n^3(c - 1/\bar{l})(1/l - 1/\bar{l})^2}{n(c - 1/l)}\bar{y}^3y\Delta\left(\frac{c}{n}\right). \tag{8.35}$$

The total Seidel or primary aberration is the sum of the five terms given by eqns (8.27), (8.29), (8.31), (8.33) and (8.35). However, these expressions are still not in very convenient forms. We see that we have actually obtained the primary aberrations for the chosen conjugates as specified by values of l, l', η etc. in each space, for the chosen pupil position as specified by the values of $\bar{l}$, $\bar{l}'$ in each space, and for a point in the exit pupil at which a particular ray emerges; this is the ray given by x, y at each surface, and it is one ray of the pencil from the off-axis object point at a distance η_0 from the axis in object space. However, the primary aberrations have known simple forms, so that if we obtain the aberration for one ray of the pencil we know it for all of the rays; e.g. primary spherical aberration varies as $(x^2 + y^2)^2$ so that if we have one wavelength of aberration at half pupil radius we must have 16 wavelengths at the rim of the pupil; similarly distortion varies as the cube of the field angle so that if there is 0·05 mm distortion at half the full field there will be 0·4 mm at full field. Thus it is sufficient to write the formulae for one point in the pupil and one field angle, suppressing the explicit dependence on pupil and field variables. The values chosen are for the full aperture in the meridian (y) section and the full field; these are respectively given by the values of h obtained from a paraxial raytrace at full aperture from the axial object point and the values of η implied by the value of the Lagrange invariant H for full aperture and field. Thus in eqn (8.27) we put $x = 0$ and $y = h$, take h^2 with $n^2(c - 1/l)^2$ to give A^2, take h with $1/nl$ to give $-u/n$ and we find for the spherical aberration sum

$$-\frac{1}{8} \sum_{\substack{\text{all} \\ \text{surfaces}}} A^2 h\Delta\left(\frac{u}{n}\right).$$

(8.36)

Similarly in eqn (8.29) the quantity $\bar{y}$ goes with $n(c - 1/\bar{l})$ to give $\bar{A}$, the refraction invariant of the principal ray, and we obtain for the coma sum

$$-\frac{1}{2} \sum_{\substack{\text{all} \\ \text{surfaces}}} \bar{A} A h\Delta\left(\frac{u}{n}\right).$$

(8.37)

The third sum, eqn (8.31), gives in the same way

$$-\frac{1}{2} \sum \bar{A}^2 h\Delta\left(\frac{u}{n}\right).$$

(8.38)

The fourth, eqn (8.33), gives

$$-\frac{1}{4} \sum \bar{A}^2 h\Delta\left(\frac{u}{n}\right) - \frac{1}{4} H^2 \sum c\Delta\left(\frac{1}{n}\right)$$

(8.39)

and the fifth, eqn (8.35), gives

$$-\frac{1}{2}\sum\frac{\overline{A}}{A}\left\{\overline{A}^2h\Delta\left(\frac{u}{n}\right) + H^2c\Delta\left(\frac{1}{n}\right)\right\}. \tag{8.40}$$

It is customary to denote these sums, with numerical factors omitted, and with some redistribution between the third and fourth terms, by single symbols, as follows:

$$
\left.
\begin{aligned}
S_{\mathrm{I}} &= -\sum A^2h\Delta\left(\frac{u}{n}\right) \\[2ex]
S_{\mathrm{II}} &= -\sum \overline{A}Ah\Delta\left(\frac{u}{n}\right) \\[2ex]
S_{\mathrm{III}} &= -\sum \overline{A}^2h\Delta\left(\frac{u}{n}\right) \\[2ex]
S_{\mathrm{IV}} &= -\sum H^2c\Delta\left(\frac{1}{n}\right) \\[2ex]
S_{\mathrm{V}} &= -\sum \left\{\frac{\overline{A}^3}{A}h\Delta\left(\frac{u}{n}\right) + \frac{\overline{A}}{A}H^2c\Delta\left(\frac{1}{n}\right)\right\}
\end{aligned}
\right\} \tag{8.41}
$$

These are the five *Seidel sums*. Now let h_{p} be the incidence height at the exit pupil of the paraxial ray from the axial object point and let $\eta_{\max}$ be the maximum image height, i.e. that implied in the value of H. Then if x_{p} and y_{p} are coordinates of any point in the exit pupil and η is any other image height it follows from eqn (8.25) that the primary wavefront aberration for x_{p}, y_{p} and η is

$$
\begin{aligned}
W(x_{\mathrm{p}}y_{\mathrm{p}}\eta) = {}& \tfrac{1}{8}S_{\mathrm{I}}\frac{(x_{\mathrm{p}}^2 + y_{\mathrm{p}}^2)^2}{h_{\mathrm{p}}^4} \\[2ex]
& + \tfrac{1}{2}S_{\mathrm{II}}\frac{y_{\mathrm{p}}(x_{\mathrm{p}}^2 + y_{\mathrm{p}}^2)}{h_{\mathrm{p}}^3}\cdot\frac{\eta}{\eta_{\max}} + \tfrac{1}{2}S_{\mathrm{III}}\frac{y_{\mathrm{p}}^2}{h_{\mathrm{p}}^2}\cdot\frac{\eta^2}{\eta_{\max}^2} \\[2ex]
& + \tfrac{1}{4}(S_{\mathrm{III}} + S_{\mathrm{IV}})\frac{(x_{\mathrm{p}}^2 + y_{\mathrm{p}}^2)}{h_{\mathrm{p}}^2}\frac{\eta^2}{\eta_{\max}^2} + \tfrac{1}{2}S_{\mathrm{V}}\frac{y_{\mathrm{p}}}{h_{\mathrm{p}}}\cdot\frac{\eta^3}{\eta_{\max}^3}. \tag{8.42}
\end{aligned}
$$

Comparing these terms with those of eqn (7.37) and the discussion which follows, in Section 7.7, we see that they do in fact represent respectively spherical aberration, coma, astigmatism, field curvature and distortion, since they have the correct dependence on the aperture and field variables. We have thus obtained the primary aberrations of the system in terms of the constructional parameters and of the data of two paraxial rays; these are the ray from the axial object point through the rim of the pupil and the principal ray from the extreme field point.

Equations (8.41) and (8.42) are substantially in the original form derived by L. Seidel in 1856 although the derivation is different. Many different versions have been given by different authors but they all yield essentially the same quantities, the five Seidel sums, except that sometimes simple numerical factors or powers of the Lagrange invariant occur; the different formulae were derived for computing convenience. For most practical optical design purposes it is sufficient to compute the Seidel sums (eqn (8.41)) and we have included eqn (8.42) simply to make explicit their significance; in practice one would always know how to obtain the aberration for fractional aperture and field positions by applying the appropriate powers of the fractions to each Seidel sum and it would not be necessary to use eqn (8.42) by calculating exit pupil coordinates.

We have grouped the terms in eqn (8.42) to correspond to those of eqn (7.37) so that the five primary aberrations can be identified. It can be seen that, quite simply, when S_I, S_{II} or S_{III} vanish then there will be no primary spherical aberration, coma or astigmatism respectively. Thus for zero S_{III} the value of S_{IV} is a measure of field curvature; however, if S_{III} is non-zero both tangential and sagittal focal surfaces (see Section 7.7) have different curvatures from that which would be given by S_{IV} alone. We can see this more clearly by regrouping the third and fourth terms of eqn (8.42) in the form

$$\tfrac{1}{4}(3S_{III} + S_{IV})\frac{y_p^2}{h_p^2}\cdot\frac{\eta^2}{\eta_{max}^2} + \tfrac{1}{4}(S_{III} + S_{IV})\frac{x_p^2}{h_p^2}\cdot\frac{\eta^2}{\eta_{max}^2}; \qquad (8.43)$$

these could then be called tangential and sagittal field curvature terms; they correspond to the focal shifts in the y- and x-sections which produce the astigmatic image surfaces indicated in Fig. 7.27. Their significance will be seen more clearly in Section 8.4 when the ray formulae for the Seidel aberrations are summarized.

The term S_{IV} is called the Petzval sum, after J. Petzval who discovered in 1843 that $S_{IV} = 0$ is the condition for a flat image field; however, it is not clear whether he understood the need for a restriction to zero astigmatism. The other four Seidel sums can be seen from eqn (8.41) to depend on the Gaussian conjugates, since A, h and u occur in the formulae, and S_{II}, S_{III} and S_V also depend on the stop position, since $\bar{A}$ occurs in the formulae; however, it can be seen that S_{IV} depends only on the curvatures and refractive indices. Thus the Petzval sum can be regarded as intrinsic to an optical system, whereas the other Seidel aberrations can vary, within limits, with conjugates and stop position.

8.3 Validity of the Seidel sum formulae

We do not know precisely to what extent the expansion of the wavefront aberration suggested in eqn (7.37) is valid in the sense of converging as a power series in the three variables. It is certain that it can only be valid for a restricted range of the variables, since discontinuities must occur for very large apertures or field angles; this happens when a ray misses a surface completely or when it is incident at an angle greater than the critical angle, as in Fig. 8.3. An interesting discussion of this matter was given by T. Y. Baker and L. N. G. Filon (1921: *Phil. Trans. R. Soc.* **221**, 29–71) and they showed that it is even possible that at a particular point in field and aperture the aberration expansion may converge if expressed in terms of one set of variables and diverge for another set; these might be, for example, the incidence height or the sine of the angle subtended at the centre of curvature by the point of incidence. A few numerical comparisons have been made between calculated higher order aberration terms and finite raytracing and these suggest that if the series is convergent it converges rather slowly for some actual optical systems. However, we may adopt a quasi-experimental approach and state that Gaussian optics gives the correct answers when used sensibly, i.e. for very small apertures and field angles; likewise the primary aberration formulae seem to give the right values at apertures and field angles immediately beyond the Gaussian region. This is true both for tests by raytracing and by experiment with lenses, but we cannot design a system which has no higher order aberrations at all so it is not possible to say conclusively that eqn (7.37) converges. However, although it is easy to construct examples of power series which diverge for all non-zero values of the variables it is generally accepted that a series obtained as a solution of a physical problem will

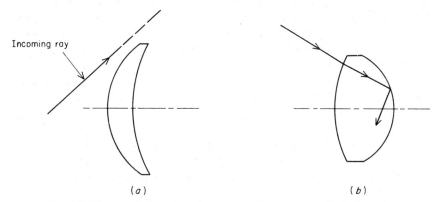

Incoming ray

(a)   (b)

FIG. 8.3. Divergence of an aberration power series expansion; (a) the incoming ray misses the surface, (b) the ray is totally internally reflected

have at least a limited range of convergence; we therefore assume that the aberration expansion is valid for ranges of the variables within the limits suggested by Fig. 8.3. The best available justification for this assumption in relation to the Seidel aberrations is, as implied above, that it seems to be correct experimentally.

Many different derivations of the Seidel sum formulae are to be found in the literature; of these the simplest to follow are perhaps those which start by deriving the spherical aberration term alone by considering an axial pencil and then use the concept of an "auxiliary axis" through the centre of curvature of the current surface to obtain the oblique aberration terms (see, for example, L. C. Martin and W. T. Welford, "Technical Optics", Vol. I, second edition (1966), Pitman, London). A derivation on these lines is shorter and more direct than that given in Section 8.2 above but is open to the objection that when the Gaussian object (or image) plane happens to lie at or near the centre of curvature of a refracting surface the auxiliary axis is almost at right-angles to the optical axis. The particular derivation used in Section 8.2 was chosen because it avoids this and similar weaknesses for particular cases (e.g. plane refracting surface, pupil at infinity, etc.). The formulae are of course equally valid in all these cases—it is merely the derivations which become suspect. In the derivation of Section 8.2 the main doubtful case is that of an object (or image) in one of the spaces being at infinity; this can easily be dealt with as a limiting case, for we have to take the difference between the expressions on the right-hand side of eqns (8.13) and (8.14), and if this be done *before* l is allowed to tend to infinity there is no difficulty.

The remaining objection to the present derivation is concerned with the expression for S_V in eqn (8.41). This or a very similar expression is almost invariably given because it shows clearly the relationship of S_V to the other four sums, but it can be seen that if the refraction invariant vanishes at one of the surfaces, i.e. if the Gaussian image plane is at the centre of curvature, the expression for S_V for that surface becomes indeterminate. This is not a very serious matter because in deriving that form of S_V we had deliberately to multiply and divide by A to get eqn (8.35). We can therefore obtain an alternative form which is never indeterminate by returning to the second line of eqn (8.35), which gives the expression:

$$\tfrac{1}{2} \sum \bar{A}\Delta(-h\bar{u}^2 - \bar{h}^2 cu) \tag{8.44}$$

This could be used as it stands, removing the factor $\tfrac{1}{2}$ to get an expression for S_V, or we can get several alternatives by expressing $\bar{u}$ and u in terms of $\bar{A}$ and A by means of eqn (6.2). The complete elimination leads to the form

$$-\bar{A}^3 h\Delta\left(\frac{1}{n^2}\right) + \bar{h}\bar{A}(2h\bar{A} - \bar{h}A)c\Delta\left(\frac{1}{n}\right); \tag{8.45}$$

this can be used to substitute for any of the terms in S_V which becomes indeterminate because the image falls at the centre of curvature.

The above discussion illustrates the fact that the Seidel sum formulae can be put in many alternative forms, as mentioned in Section 8.2. As with finite raytracing the main reason for the development of many alternatives was computing convenience when electronic computers were not available; nowadays it is appropriate to use the forms which show clearly the interrelations between the aberrations and their dependence on the constructional data etc. There seems little doubt that Seidel's original forms, i.e. eqn (8.42), do this best.

8.4 Ray aberration expressions from the Seidel sums

As explained in Section 8.2 the Seidel sums represent, apart from simple numerical factors, the wavefront aberration for each of the primary aberration at full aperture and field. They arise naturally in this form because the summation theorem for aberrations applies specifically to optical path terms. However, it is also convenient to have ray aberration expressions and in this section we apply eqns (7.15) and related results to obtain such expressions.

Thus for spherical aberration we have, from the first term of eqn (8.42), the expression

$$\tfrac{1}{8}S_I\left(\frac{r_p}{h_p}\right)^4, \tag{8.46}$$

where r_p is the radius in the exit pupil at which the aberration is to be found; applying eqn (7.15) the transverse ray aberration is

$$-\frac{R}{n}\cdot\frac{S_I}{2h_p^4}\,r_p^3 \tag{8.47}$$

and since $h_p/R = -u$, the convergence angle, we have for the transverse ray aberration at full aperture and at the Gaussian image plane

$$\frac{S_I}{2nu}. \tag{8.48}$$

The longitudinal spherical aberration, i.e. the distance from the Gaussian image point at which the ray from the edge of the pupil meets the axis, is obtained by dividing this quantity by the convergence angle:

$$\frac{S_I}{2nu^2}, \tag{8.49}$$

here we take the longitudinal aberration as positive if the ray intersection is to the left of the Gaussian image point, the usual convention.

The results for the other aberrations will be stated without detailed explanation since the procedure for obtaining them is very similar to that used above. In the geometrical coma pattern, Fig. 8.4, the principal ray meets the image plane at P_1, the two rays from the sagittal section meet at A and the

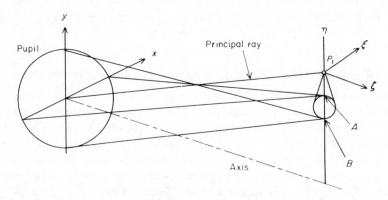

FIG. 8.4. Sagittal and tangential coma

rays from the tangential (meridian) section meet at B, as shown in Section 7.7. The distances P_1A and P_1B are called the sagittal and tangential coma respectively and for them we have

$$\text{Sagittal coma} = \frac{S_{II}}{2nu}, \qquad (8.50)$$

$$\text{Tangential coma} = \frac{3S_{II}}{2nu}. \qquad (8.51)$$

The three to one ratio between tangential and sagittal coma can also be deduced from the geometry of Fig. 7.20.

The third and fourth terms in eqn (8.42) represent astigmatism and field curvature respectively; thus the length of either of the astigmatic focal lines must be twice the transverse ray aberration for the third term, i.e.

$$\text{Length of focal line} = \left| \frac{2S_{III}}{nu} \right|, \qquad (8.52)$$

which agrees with the result obtained in Section 7.7 if we make the correct substitution for the aberration coefficient b_3. Similarly we find the following:

$$\text{Distance between focal lines} = \frac{S_{\text{III}}}{nu^2}. \tag{8.53}$$

$$\begin{matrix}\text{Distance from Gaussian image plane} \\ \text{to tangential focal line}\end{matrix} = \frac{3S_{\text{III}} + S_{\text{IV}}}{2nu^2}. \tag{8.54}$$

$$\begin{matrix}\text{Distance from Gaussian image plane} \\ \text{to sagittal focal line}\end{matrix} = \frac{S_{\text{III}} + S_{\text{IV}}}{2nu^2}. \tag{8.55}$$

The latter two are taken positive if the focal lines are to the left of the Gaussian image plane, like similar quantities given below.

$$\begin{matrix}\text{Distance from Gaussian image plane to the} \\ \text{astigmatic disc of least confusion}\end{matrix} = \frac{2S_{\text{III}} + S_{\text{IV}}}{2nu^2}. \tag{8.56}$$

The Petzval sum S_{IV} only gives a realistic measure of field curvature if there is no astigmatism but it is, as explained in Section 8.3, of fundamental significance in aberration theory since it does not depend on conjugates or stop position; it is therefore customary to speak of the "Petzval surface" as an image surface near the Gaussian image plane even when there is astigmatism in the system and we have,

$$\text{Distance from Gaussian image surface to Petzval surface} = \frac{S_{\text{IV}}}{2nu^2}. \tag{8.57}$$

The four image surfaces implicitly defined by eqns (8.54) to (8.57) are, in the approximation to which the Seidel aberrations are calculated, spherical and they can be represented as in Fig. 8.5. It can be seen from the formulae that in fact only two independent quantities are involved, say S_{III} and S_{IV}; from this it follows that the Petzval curvature and astigmatism can separately have any given values, so that one could have astigmatism acting in effect as a field-flattening influence, as in Fig. 8.6; however, the distance of the tangential image surface from the Petzval surface must always be three times the distance of the sagittal image surface from the Petzval surface. The actual curvatures of these surfaces can easily be obtained and they are occasionally useful:

$$\text{Curvature of tangential image surface} = \frac{n(3S_{\text{III}} + S_{\text{IV}})}{H^2}. \tag{8.58}$$

$$\text{Curvature of sagittal image surface} = \frac{n(S_{\text{III}} + S_{\text{IV}})}{H^2}. \tag{8.59}$$

$$\begin{matrix}\text{Curvature of surface containing disc} \\ \text{of least confusion}\end{matrix} = \frac{n(2S_{\text{III}} + S_{\text{IV}})}{H^2}. \tag{8.60}$$

$$\text{Curvature of Petzval surface} = \frac{nS_{\text{IV}}}{H^2}. \tag{8.61}$$

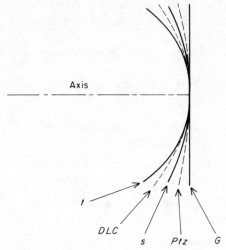

FIG. 8.5. The image surfaces; G—Gaussian, t—tangential, s—sagittal, Ptz—Petzval, DLC—disc of least confusion

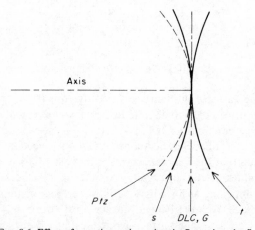

FIG. 8.6. Effect of negative astigmatism in flattening the field

Distortion is expressed as a ray aberration in several different ways. The actual transverse ray aberration of the principal ray, i.e. the displacement of the image point, is

$$\text{Distortion} = \delta\eta' = \frac{S_V}{2nu}. \tag{8.62}$$

If this is written as a fraction of the ideal image height we obtain

$$\text{Fractional distortion} = \frac{\delta\eta'}{\eta'} = \frac{S_V}{2H}. \tag{8.63}$$

We can also calculate the rate of change of image height with respect to object height, which is, in effect, the local magnification in the tangential (meridian) section. We find

$$\text{Differential distortion} = \frac{d\eta'}{d\eta} = m\left(1 + \frac{3S_V}{2H}\right) \tag{8.64}$$

In all the expressions in this section, it is, of course, implied that n and u refer to the space in which the aberrations are to be computed, usually the final image space. These expressions all break down if this happens to be a star space, i.e. if $u = 0$ and in that case there are three alternatives: (a) to keep to the wavefront aberrations, i.e. the Seidel sums themselves, which are meaningful in star space as elsewhere; (b) to add a notional perfect optical system with non-zero power, so that a finite conjugate is obtained; or (c) to use angular ray aberrations. The third is sometimes useful and we give the relevant expressions below, referred as usual to full field and aperture; they are all obtained by applying eqns (7.15) with both sides divided by R.

$$\text{Angular spherical aberration} = -\frac{S_I}{2nh_p} = \frac{\beta S_I}{2H} \tag{8.65}$$

where β is the field angle in the star space.

$$\text{Angular tangential coma} = -\frac{3S_{II}}{2nh_p} = \frac{3\beta S_{II}}{2H} \tag{8.66}$$

$$\text{Angular sagittal coma} = -\frac{S_{II}}{2nh_p} = \frac{\beta S_{II}}{2H} \tag{8.67}$$

$$\text{Angular distortion} = -\frac{S_V}{2nh_p} = \frac{\beta S_V}{2H} \tag{8.68}$$

The concept of angular aberration is a trifle artificial when applied to astigmatism and field curvature. We can incorporate these in the scheme by using the angular aberration of the rays in the two sections with respect to the Gaussian image point, so that we obtain two terms corresponding to the terms in eqn (8.43), i.e. to eqns (8.54) and (8.55). We then have,

$$\text{Angular tangential astigmatism} = -\frac{(3S_{III} + S_{IV})}{2nh_p}$$

$$= \frac{\beta(3S_{III} + S_{IV})}{2H}. \tag{8.69}$$

$$\text{Angular sagittal astigmatism} = -\frac{(S_{III} + S_{IV})}{2nh_p}$$

$$= \frac{\beta(S_{III} + S_{IV})}{2H} \tag{8.70}$$

It should be noted that these angular aberration formulae are not very suitable for use unless the image *is* in a star space because it is only in this case that the values of β and h_p which appear are actually independent of the location of the reference sphere. If the pupil were telecentric (which could not, of course, happen in a star space since the pupil and image would coincide) the angular aberration expressions would be quite meaningless.

The choice of the appropriate measure of aberration depends on the nature of the problem. Wavefront aberrations arise naturally in the development of the subject because of the summation theorem for primary aberrations but in discussing distortion it is almost invariably thought of in terms of transverse ray aberration and astigmatism is frequently considered as a longitudinal ray aberration, as in eqns (8.53) to (8.56). Frequently the complete system of Seidel aberration formulae is presented as surface-by-surface summations to give longitudinal or transverse ray aberrations (e.g. A. E. Conrady (1929), "Applied Optics and Optical design", Part I, Oxford University Press) but in fact these summations are really of wavefront aberration terms and the conversion to ray aberrations is done after the summation by formulae equivalent to those given here.

8.5   Computation of the Seidel sums; effect of stop shifts

As with finite raytracing, much work has been done in transforming the Seidel sum formulae to obtain marginal gains in computing speed, but this has lost its point with the availability of fast electronic computers and it is necessary only to ensure that the formulae used are such that no losses in significant figures occur for the required ranges of values of the input data. It is easily shown that this is the case for all the expressions in eqn (8.41) with the exception of S_V and this can easily be included by using eqn (8.44), eqn (8.45) or many easily obtained alternatives. Thus the data required are, besides the curvatures, indices and separations of the system, the incidence heights, convergence angles and refraction invariants of the paraxial ray from the axial object point through the rim of the aperture stop, together with the refraction invariants of the paraxial principal ray from the edge of the field, and, of course, the Lagrange invariant. The latter is, as mentioned in Chapter 6, a link between field and aperture, enabling either to be expressed in terms of the other. The refraction invariants of the principal ray involve implicitly the pupil position in every space and, through the Lagrange invariant, the field angle. Thus all the data are available by simply tracing the above mentioned two paraxial rays, and quite frequently this is all that is done.

However, there is an alternative approach to the computation in which only the paraxial ray from the axial object point is traced and the refraction invariants of the principal ray are obtained from these data together with a statement of the aperture stop position and the Lagrange invariant. This offers no direct advantage as a numerical technique but the formulae obtained are of great importance because they show the dependence of the Seidel aberrations on the stop position and they permit a simple numerical calculation of these effects. The main results to be used, the Seidel difference formulae, were derived in Section 6.3; we defined the eccentricity parameter $E = \bar{h}/(hH)$ and we showed that the increment in E from surface to surface is given, with an extension of our present notation, by

$$\Delta E = \frac{-d'}{n'hh_{+1}} \tag{8.71}$$

(cf. eqn (6.16)). We can set a ghost surface at the aperture stop and E must be identically zero there, so that eqn (8.71) enables us to compute E at all surfaces. Then eqn (6.9) gives the values of $\bar{A}$ at each surface immediately and we then have all that is needed for computing the Seidel sums. This method has some slight advantage of speed for numerical work over the method of simply tracing the two paraxial rays, but its real advantage is in the insight it yields into the effects of changes of the axial position of the aperture stop, as we shall now show.†

We know from the Seidel sum formulae that S_{II}, S_{III} and S_V depend on the stop position, since the refraction invariant of the principal ray appears in them; thus we may expect that changes in coma, astigmatism and distortion can be made by moving the stop along the optical axis; clearly this could be a useful design operation, so we investigate its possibilities. Suppose that the stop is between surfaces k and $k + 1$ as in Fig. 8.7 and let it be moved a distance z to the right, changing its diameter at the same time in such a way that

---

† The Seidel difference formulae contravene a principle we have been applying, that all expressions should be of uniform accuracy, i.e. they must not lose numerical precision or become indeterminate for certain ranges of the variables. Clearly E tends to infinity at a field surface, since h tends to zero, and under this condition (6.9) for $\bar{A}$ is indeterminate. This situation is very rare because there are important practical disadvantages in having an image at or very near to a refracting surface, but for completeness we give some expressions which circumvent the difficulty. If surface k is the field surface then it is easily shown from eqns (6.16) and (6.9) that

$$\bar{A}_k = HA_k E_{k-1} - \frac{H}{h_{k-1}}(1 + d'_{k-1}c_k)$$

$$= HA_k E_{k+1} - \frac{H}{h_{k+1}}(1 - d'_k c_k).$$

The first equation yields $\bar{A}_k$ without indeterminacy and both together give a bridging relation between E_{k-1} and E_{k+1}.

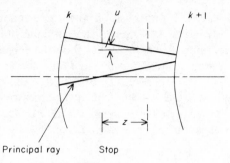

FIG. 8.7. Shifting the aperture stop

the size of the axial pencil is unchanged. It can be seen that the effect is to increase the value of the eccentricity E at surface $k + 1$ by

$$\delta E = \frac{z}{n_s h_s (h_s + z u_s)}, \tag{8.72}$$

where h_s is the radius of the stop at its original position, u_s is the convergence angle of the ray from the centre of the object and n_s is the refractive index in the stop space. But, as shown in Chapter 6, E is obtained by summing increments in E for each space, so that δE in eqn (8.72) is the increment in E at *every* surface of the system, corresponding to the shift z in the stop space.

It is therefore convenient to express the results of a shift of the stop in terms of δE, rather than the actual value of the axial movement of the stop, since δE refers equally to every space in the system whereas the axial movement of the pupil varies according to the longitudinal magnification of the pupil. In what follows we shall always use δE as the stop shift parameter and it will be understood that the actual axial movement is found from eqn (8.72).

In order to find the changes in the Seidel sums due to a stop shift δE we note first that from eqn (6.9) the value of $\bar{A}$ at any surface becomes

$$\bar{A} + \delta \bar{A} = \bar{A} + H \, \delta E . A; \tag{8.73}$$

if this is substituted for $\bar{A}$ in eqn (8.41) we obtain for the aberration changes

$$\left. \begin{aligned} \delta S_{II} &= \varepsilon S_I \\ \delta S_{III} &= 2\varepsilon S_{II} + \varepsilon^2 S_I \\ \delta S_V &= \varepsilon(3 S_{III} + S_{IV}) + 3\varepsilon^2 S_{II} + \varepsilon^3 S_I \end{aligned} \right\}, \tag{8.74}$$

where we have put ε for $H \, \delta E$. As noted above S_I and S_{IV} do not depend on the stop position. These formulae, generally called the *stop-shift formulae*, are of considerable importance in aberration theory and in practical optical design.

It is of interest to see in a graphical way how a movement of the stop can, as it were, transform one aberration into another. Let Σ in Fig. 8.8 be a wavefront from an off-axis object point with, as indicated, spherical aberration; if a small stop is placed so as to produce an exit pupil as indicated in the diagram the effect will be to select the central part only of the aberrated wavefront, so producing a symmetrical image; if the stop is moved along the axis the pupil will move, to the left, say, and the effect will be to select a different

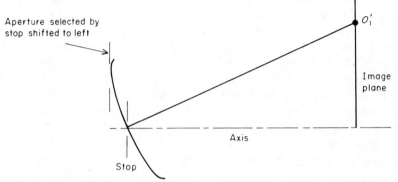

FIG. 8.8. Effect of axial stop shift in converting spherical aberration to coma and astigmatism

portion of the wavefront, as indicated; but if the original wavefront had spherical aberration this selected region will be more strongly curved at the upper half than in the lower half and it will thus be asymmetric in a coma-like way. A deeper study shows also that astigmatism and distortion are introduced. The same thing can be seen algebraically: spherical aberration is represented by

$$W = b_1(x^2 + y^2)^2 \tag{8.75}$$

and from Fig. 8.7 a stop shift is equivalent to a change of origin to, say, $(0, y_0)$; the resultant expression for W becomes

$$W = b_1[(x^2 + y^2)^2 + 4y_0y(x^2 + y^2) + 6y_0^2y^2 + 2y_0^2x^2 + 4y_0^3y]. \tag{8.76}$$

The additional terms can be identified as corresponding to coma, tangential field curvature, sagittal field curvature and distortion.

In the above discussion we spoke of a "small" stop and it can be shown from the formulae that the most marked stop-shift effects occur with a small stop and large field angles; again resorting to the graphical illustration, the pupil in Fig. 8.8 could be so small that the spherical aberration of the wavefront transmitted is quite negligible but yet when the stop is moved far enough the displaced pupil can select a very asymmetric portion of the wavefront.

Under these conditions of large field and small aperture the formulae suggest that the stop position can be an important variable; for example, if S_I is not zero the stop position can be chosen by the first of eqns (8.74) to give zero S_{II} and again by the second equation S_{III} can be made zero for non-zero S_{II}, and so on. Conversely, S_{II} is independent of the stop position if S_I is zero, S_{III} is independent of stop position if S_{II} and S_I are zero and similarly for S_V. There are, however, some important practical limitations to these elegant theorems. First it should be noted that they apply only to the primary aberrations; higher order aberrations are also, of course, affected by stop shifts, but in a much more complicated way, and it is in practice necessary to use ray-tracing to determine such effects. Secondly it may happen that the amount of stop shift required to produce a certain desired effect, as indicated by the value of $H\,\delta E$, may put the stop into a physically inconvenient or even un-realizable position, say inside a piece of glass or in front of the object. Finally the effect of the stop in controlling aberrations may be considerably modified by vignetting at the rim of one or more elements; such vignetting is sometimes deliberately allowed in order to eliminate badly aberrated parts of the extreme off-axis wavefronts. With all these reservations it is still true that stop-shift effects are of great importance in optical design.

8.6 Aspheric surfaces

To the approximation used in Seidel theory a spherical surface of curvature c has the equation

$$z = \tfrac{1}{2}c(x^2 + y^2) + \tfrac{1}{8}c^3(x^2 + y^2)^2 + \cdots. \tag{8.77}$$

The effect of aspherizing this surface in axisymmetric fashion, while keeping the axial curvature unchanged, can clearly be specified by just one parameter, the coefficient G of an additional fourth degree term $G(x^2 + y^2)^2$. Other differences between possible aspherics and the sphere such as are discussed in Sections 4.7 and 4.8 affect only the 6th and higher degree terms and so they cannot affect the Seidel aberrations. It is therefore convenient to specify an aspheric in Seidel approximation by the addition of a *figuring term*, as it is called, to the spherical shape:

$$z = \tfrac{1}{2}c(x^2 + y^2) + (\tfrac{1}{8}c^3 + G)(x^2 + y^2)^2 + \cdots. \tag{8.78}$$

The expression "figuring" derives from one process of manufacture in which the surface is first polished to a spherical shape and then retouched or figured to a shape a few wavelengths different from the sphere and it thus suggests that the surface departs only slightly from the spherical shape, but we shall call G the figuring coefficient or aspheric coefficient irrespective of the magnitude of the asphericity.

We could have included the effect of aspherizing in the original derivation of the Seidel sums (Section 8.2) by using eqn (8.78) as the equation of the refracting surface, but there would have been rather more complication in the calculation; since in practice it is very unusual to have more than two aspheric surfaces in any optical system it seems simpler to take account of aspherizing by a special treatment, as follows.

The aspheric term affects only the fourth power of the aperture; thus it can be seen that if the aperture stop or pupil is at an aspheric surface the only effect of the aspherizing is to change the spherical aberration, since all other Seidel aberrations depend on lower powers than the fourth in the aperture. To be precise, the wavefront at (x, y) in the pupil is advanced relative to the centre by $(n' - n)G(x^2 + y^2)^2$ and so the effect of the figuring is to add to S_1 a term

$$S_1^* = \sum 8Gh^4\Delta(n) \tag{8.79}$$

where the summation is taken over all surfaces which are aspherized and the appropriate figuring coefficient G is used at each.

If an aspherized surface is not at or conjugate to the aperture stop the aspherizing will introduce coma, astigmatism and distortion, by the same reasoning which was used in connection with Fig. 8.7 to show how a shift of stop produces these aberrations at a surface where there is spherical aberration. Thus at the aspheric surface there will be the following increments of the oblique aberrations

$$\left.\begin{aligned} S_{II}^* &= \varepsilon S_I^* \\ S_{III}^* &= \varepsilon^2 S_I^* \\ S_V^* &= \varepsilon^3 S_I^* \end{aligned}\right\}. \tag{8.80}$$

These are increments to the aberrations at the aspheric surface only and they are given in terms of the aspheric contribution S_I^* at that particular surface. If there is more than one aspheric each one has an aspheric contribution S_I^* and each of these gives rise to coma, astigmatism and distortion contributions according to eqns (8.80) *at its own surface*.

8.7 Effect of change of conjugates on the primary aberrations

The object and image conjugates appear in the Seidel sums through the quantities u, h, etc. in eqns (8.41) and similarly in the other expressions for primary aberrations given in this chapter, so that the dependence of the aberrations on the conjugates is in a sense already given; equally the dependence on the stop position is given through the value of $\bar{A}$ which appears in these equations. However, we found in Section 8.5 that it is possible to obtain relatively simple expressions for the effect of a *change* of the stop position on the stop-shift formulae, and these have numerous applications; in particular they are useful in the important thin lens stage of design (Chapter 12). It turns out to be possible to obtain similar formulae for the effect of a change of conjugates and we shall derive these in this section. The concept of a change of conjugates is not of such basic importance in optical design proper as movement of the aperture stop since optical systems are usually designed for prescribed fixed conjugates, although many systems are in practice used over a range of conjugates; also the formulae lead to several interesting results on the questions of the possibility of complete correction of primary aberrations. We follow mainly the derivation due to C. G. Wynne.[†]

The principle of the method is to define aberrations of *pupil* imagery which are analogous to the ordinary Seidel sums for object imagery and to apply to these pupil aberrations formulae for "stop-shifts"; since the "pupil" for pupil imagery is in fact the image plane we shall then obtain expressions for the effects of conjugate shifts on the pupil aberrations. But the pupil aberrations are, as it turns out, very simply related to the Seidel sums and so by substitution we arrive at the effects of the conjugate shift on the Seidel sums. The substitution and reductions are laborious, but not intrinsically difficult, and we shall indicate the methods only briefly. We define first the pupil aberration terms:

$$
\left.
\begin{aligned}
\bar{S}_{\mathrm{I}} &= \bar{A}^2 \bar{h} \Delta\left(-\frac{\bar{u}}{n}\right) \\[6pt]
\bar{S}_{\mathrm{II}} &= \frac{A}{\bar{A}}\, \bar{S}_{\mathrm{I}} \\[6pt]
\bar{S}_{\mathrm{III}} &= \frac{A}{\bar{A}}\, \bar{S}_{\mathrm{II}} \\[6pt]
\bar{S}_{\mathrm{IV}} &= S_{\mathrm{IV}} = H^2 c \Delta\left(-\frac{1}{n}\right) \\[6pt]
\bar{S}_{\mathrm{V}} &= \frac{A}{\bar{A}}\,(\bar{S}_{\mathrm{III}} + S_{\mathrm{IV}}).
\end{aligned}
\right\}
\tag{8.81}
$$

These are defined exactly as for the Seidel aberration terms but with the roles of pupil and image plane interchanged. We also define $\bar{H} = -H$ and $\bar{H}\bar{E} = h/\bar{h} = (HE)^{-1}$ and we recall from Chapter 6 certain relations:

$$
\left.
\begin{aligned}
A &= n(hc + u) \\[4pt]
H &= n(u\bar{h} - \bar{u}h) \\[4pt]
&= A\bar{h} - \bar{A}h \\[4pt]
\bar{A} &= -\frac{H}{h}(1 + AhE)
\end{aligned}
\right\}.
\tag{8.82}
$$

† *Proc. phys. Soc.* **65B**, 429–437 (1952).

We next derive the expressions for the pupil aberrations in terms of the Seidel sums. We have

$$\bar{S}_I = \bar{A}^2 \bar{h} \Delta\left(-\frac{\bar{u}}{n}\right)$$

$$= HE.\bar{A}^2 h \Delta\left(\frac{\bar{h}c}{n} - \frac{\bar{A}}{n^2}\right)$$

$$= HE\left\{\frac{\bar{A}^3}{A} h \Delta\left(-\frac{(hc + u)}{n}\right) + \bar{A}^2 h \bar{h} c \Delta\left(\frac{1}{n}\right)\right\}$$

$$= HE\left\{\frac{\bar{A}^3}{A} h \Delta\left(-\frac{u}{n}\right) - \left(\frac{\bar{A}^3}{A} h^2 c - \bar{A}^2 h \bar{h} c\right)\Delta\left(\frac{1}{n}\right)\right\}$$

$$= HE\left\{\frac{\bar{A}^3}{A} h \Delta\left(-\frac{u}{n}\right) - \frac{\bar{A}^2}{A} hc(\bar{A}h - A\bar{h})\Delta\left(\frac{1}{n}\right)\right\}$$

$$= HE\left\{\frac{\bar{A}^3}{A} h \Delta\left(-\frac{u}{n}\right) + \frac{\bar{A}}{A} Hhc\Delta(\bar{h}c + \bar{u})\right\}$$

$$= HE\left\{\frac{\bar{A}^3}{A} h \Delta\left(-\frac{u}{n}\right) + \frac{\bar{A}}{A} Hc\Delta(\bar{h}hc + \bar{h}u + h\bar{u} - \bar{h}u)\right\}$$

$$= HE\left\{\frac{\bar{A}^3}{A} h \Delta\left(-\frac{u}{n}\right) + \frac{\bar{A}}{A} Hc\Delta\left(\frac{\bar{h}A}{n} - \frac{H}{n}\right)\right\}.$$

Thus we have the result

$$\bar{S}_I = HE\left\{S_V + \bar{A}\bar{h}Hc\Delta\left(\frac{1}{n}\right)\right\} \tag{8.83}$$

Making similar substitutions in the expression for $\bar{S}_{II}$ in eqn (8.81) we obtain

$$\bar{S}_{II} = S_V - H\Delta(\bar{u}^2) \tag{8.84}$$

and similar results can be obtained for $\bar{S}_{III}$ and $\bar{S}_V$. Collecting them together we have

$$\left.\begin{array}{l} \bar{S}_I = HE\left\{S_V + \bar{A}\bar{h}Hc\Delta\left(\frac{1}{n}\right)\right\} \\ \bar{S}_{II} = S_V - H\Delta(\bar{u}^2) \\ \bar{S}_{III} = S_{III} - H\Delta(u\bar{u}) \\ \bar{S}_V = S_{II} - H\Delta(u^2) \end{array}\right\}. \tag{8.85}$$

It may be noted that $\bar{S}_I$, the spherical aberration of the pupil, is sometimes denoted by S_{VI} and regarded as a sixth Seidel aberration; by symmetry S_I is then alternatively called $\bar{S}_{VI}$. The justification for this is that $\bar{S}_I$ actually appears in the characteristic function development of the aberrations as the term in b_6 in eqn (7.37); for reasons explained in Section 7.6 it is not regarded as an aberration in the strict sense but we shall find that when the effects of conjugate shifts are worked out $\bar{S}_I$ makes an important contribution.

The next stage is to note that a change in conjugates can be specified by the same increment to $\overline{HE}$ at all surfaces, $\bar{H}\,\delta\bar{E}$, say; here it is assumed that as the image planes shift in the various spaces of the system the field sizes are adjusted in such a way as to keep $\bar{H}$, the Lagrange invariant, constant, just as for stop shifts it was assumed that the new pupil diameters were adjusted to keep H constant. We then have, by direct analogy with eqns (8.74),

$$\left.\begin{aligned}
\bar{S}_I^* &= \bar{S}_I \\
\bar{S}_{II}^* &= \bar{S}_{II} + \bar{\varepsilon}\bar{S}_I \\
\bar{S}_{III}^* &= \bar{S}_{III} + 2\bar{\varepsilon}\bar{S}_{II} + \bar{\varepsilon}^2\bar{S}_I \\
\bar{S}_V^* &= \bar{S}_V + \bar{\varepsilon}(3\bar{S}_{III} + S_{IV}) + 3\bar{\varepsilon}^2\bar{S}_{II} + \bar{\varepsilon}^3\bar{S}_I
\end{aligned}\right\}, \tag{8.86}$$

where we have put $\bar{\varepsilon}$ for $\bar{H}\,\delta\bar{E}$, the change in $\overline{HE}$, and where, for example, $\bar{S}_{II}^*$ denotes the value of $\bar{S}_{II}$ after the conjugate shift. It is implied that these terms are to be summed over all surfaces as usual. There is one more result needed to complete eqns (8.86), namely an expression for $\bar{S}_{VI}^*$ or S_I^*. This is not available from Section 8.5 because we did not compute the effects of stop shifts on S_{VI}. We have from Sections 8.1 and 6.3

$$\left.\begin{aligned}
S_I &= A^2 h\Delta\left(-\frac{u}{n}\right) \\
A^* &= A + \bar{\varepsilon}\bar{A} \\
h^* &= h + \bar{\varepsilon}\bar{h} \\
u^* &= u + \bar{\varepsilon}\bar{u}
\end{aligned}\right\}, \tag{8.87}$$

so that

$$S_I^* = (A + \bar{\varepsilon}\bar{A})^2(h + \bar{\varepsilon}\bar{h})\Delta\left(\frac{-(u + \bar{\varepsilon}\bar{u})}{n}\right)$$

$$= (A + \bar{\varepsilon}\bar{A})^2\left\{h\Delta\left(-\frac{u}{n}\right) + \bar{\varepsilon}^2\bar{h}\Delta\left(-\frac{\bar{u}}{n}\right) + \bar{\varepsilon}\bar{h}\Delta\left(-\frac{u}{n}\right) + \bar{\varepsilon}h\Delta\left(-\frac{\bar{u}}{n}\right)\right\}. \tag{8.88}$$

From eqn (6.6) we have

$$\left.\begin{aligned}
\frac{h}{\bar{h}} &= \frac{A}{\bar{A}} - \frac{H}{h\bar{A}} \\
\frac{\bar{h}}{h} &= \frac{\bar{A}}{A} + \frac{H}{hA}
\end{aligned}\right\}, \tag{8.89}$$

and substituting in the last two terms in the second factor of eqn (8.88) we have

$$\bar{\varepsilon}\hbar\Delta\left(-\frac{u}{n}\right) + \bar{\varepsilon}h\Delta\left(-\frac{\bar{u}}{n}\right) = \bar{\varepsilon}\frac{\bar{A}}{A}h\Delta\left(-\frac{u}{n}\right) + \bar{\varepsilon}\frac{A}{\bar{A}}\hbar\Delta\left(-\frac{\bar{u}}{n}\right)$$

$$+ \bar{\varepsilon}\frac{H}{A}\Delta\left(-\frac{u}{n}\right) - \varepsilon\frac{H}{\bar{A}}\Delta\left(-\frac{\bar{u}}{n}\right).$$

In this we put $u = A/n - hc$, $\bar{u} = \bar{A}/n - \bar{h}c$ in the last two terms on the right-hand side and substituting back in eqn (8.88) we obtain

$$S_I^* = (A + \bar{\varepsilon}\bar{A})^2 \left\{ h\Delta\left(-\frac{u}{n}\right) + \bar{\varepsilon}^2\hbar\Delta\left(-\frac{\bar{u}}{n}\right) \right.$$

$$\left. + \bar{\varepsilon}\frac{\bar{A}}{A}h\Delta\left(-\frac{u}{n}\right) + \bar{\varepsilon}\frac{A}{\bar{A}}\hbar\Delta\left(-\frac{\bar{u}}{n}\right) + \bar{\varepsilon}\frac{H^2}{A\bar{A}}c\Delta\left(-\frac{1}{n}\right) \right\}.$$

When the terms in the product are collected and compared with the definitions of the Seidel terms (eqns (8.41) and (8.81)) we obtain

$$S_I^* = S_I + \bar{\varepsilon}(3S_{II} + \bar{S}_V) + \bar{\varepsilon}^2(3S_{III} + 3\bar{S}_{III} + 2S_{IV})$$
$$+ \bar{\varepsilon}^3(3\bar{S}_{II} + S_V) + \bar{\varepsilon}^4\bar{S}_I. \tag{8.90}$$

That is the required equation to go with eqns (8.86), recalling that S_I is the same as $\bar{S}_{VI}$.

The final step in calculating the effects of conjugate shifts is to substitute the values of $\bar{S}_{II}$, $\bar{S}_{III}$ and $\bar{S}_V$ from eqns (8.85) into eqns (8.86) and (8.90); then, using also the last of eqns (8.87) we obtain

$$\left.\begin{aligned}
S_I^* &= S_I + \bar{\varepsilon}(4S_{II} - H\Delta(u^2)) + \bar{\varepsilon}^2(6S_{III} - 3H\Delta(u\bar{u}) + 2S_{IV}) \\
&\quad + \bar{\varepsilon}^3(4S_V - 3H\Delta(\bar{u}^2)) + \bar{\varepsilon}^4 S_{VI} \\
S_{II}^* &= S_{II} + \bar{\varepsilon}(3S_{III} + S_{IV} - H\Delta(u\bar{u})) + \bar{\varepsilon}^2(3S_V - 2H\Delta(\bar{u}^2)) + \bar{\varepsilon}^3 S_{VI} \\
S_{III}^* &= S_{III} + \bar{\varepsilon}(2S_V - H\Delta(\bar{u}^2)) + \bar{\varepsilon}^2 S_{VI} \\
S_V^* &= S_V + \bar{\varepsilon}S_{VI} \\
S_{VI}^* &= S_{VI}
\end{aligned}\right\} . \tag{8.91}$$

These are the required conjugate-shift equations; they are more complex than the stop-shift equations, mainly on account of the involvement of the spherical aberration of the pupil $\bar{S}_I$ ($=S_{VI}$). As for the stop-shift equations, it is understood that S_I, S_{II} etc. stand for the sums over all surfaces; terms like $\Delta(u^2)$ need only be taken between the initial and final spaces since intermediate terms cancel.

C. G. Wynne in the reference quoted above pointed out several general results which follow from eqns (8.91) and we give some here. Distortion does not change with conjugate shift if the spherical aberration of the pupil is zero. If *also* $2S_V - H \Delta(\bar{u}^2)$ vanishes then the astigmatism is invariant and since $S_{VI} = 0$ the distortion must also be invariant. This distortion would actually vanish if the pupils were at the nodal points so† that $\Delta(\bar{u}^2) = 0$.

Considering now the possibility of invariant coma, let the pupils be at the nodal points and let the system have air in the object and image spaces; then it follows that $H\Delta(u\bar{u}) = H^2K$, where K is the power of the system, and from the second of eqns (8.91) there are three conditions for the coma to be invariant with conjugates; these are

$$3S_{III} + S_{IV} - H^2K = 0, \qquad S_V = 0, \qquad S_{VI} = 0.$$

If the system has zero and invariant distortion and invariant astigmatism, the second and third conditions are satisfied; if also the astigmatism is zero, the first condition means equality of power and Petzval sum, so that the coma cannot also be invariant if the system has a flat field and non-zero power.

Since S_I depends on the fourth power of $\bar{\varepsilon}$, the conjugate shift, there could be four sets of conjugates for which the primary spherical aberration is zero. Likewise there could be three sets of conjugates for zero coma, two for astigmatism and one for distortion. C. G. Wynne considers in detail the possibilities of coincidences between these conjugates and shows that while it is impossible to have a non-trivial system corrected for all aberrations at two different conjugates, a result originally given by J. Clerk Maxwell, certain combinations of aberrations *can* be corrected at two conjugates.

8.8 Aplanatic surfaces and other aberration-free cases

The aplanatism concept and certain generalizations are discussed in detail in Section 9.4. In this section we shall apply the Seidel sum formula (eqns (8.41)) to show that spherical surfaces used at certain conjugates are aplanatic. The results are frequently applied in a wide range of optical designs.

From eqns (8.41) if $\Delta(u/n)$ vanishes then the surface makes no contribution to spherical aberration, coma or astigmatism. This condition is

$$\frac{u'}{n'} = \frac{u}{n}$$

i.e.

$$n'l' = nl.$$

† An invariant S_V could also be zero if the pupils were at the points of magnification -1, so that $\bar{u} = -\bar{u}'$, but this is an unlikely arrangement in an actual optical system.

If this is used to eliminate l or l' from the conjugate distance equation for a single surface (eqn (3.38)) we obtain

$$\left.\begin{array}{l} l = \dfrac{n + n'}{n}\, r \\[3mm] l' = \dfrac{n + n'}{n'}\, r \end{array}\right\} \qquad (8.92)$$

where r is the radius of curvature of the surface. The configuration of the incident and refracted rays is shown in Fig. 8.9. It can easily be shown that if the conjugates are as in eqn (8.92) the surface actually introduces no spherical aberration at all, i.e. it is not merely primary spherical aberration which vanishes. The transverse magnification of the system in Fig. 8.9(a) used in reverse is easily found to be $(n'/n)^2$; thus a system as in Fig. 8.10 forms a very well corrected magnifier, a fact familiar to collectors of glass paperweights. It is also used as the first stage in high power microscope objectives, where it serves the useful purpose of reducing the divergence of the rays considerably without introducing spherical aberration, astigmatism or coma.

From eqns (8.41) we note also the more obvious fact that S_I, S_{II} and S_{III} vanish if $h = 0$, i.e. if the image is formed at the surface, as in a field lens.

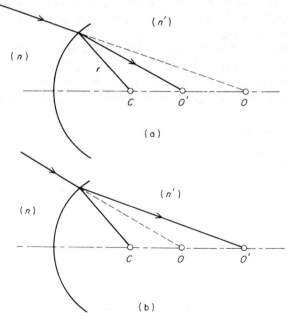

FIG. 8.9. Aplanatic refraction (a) $n' > n$, (b) $n' < n$

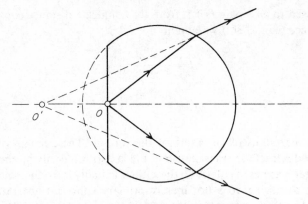

FIG. 8.10. An aplanatic magnifier

Also S_I and S_{II} vanish if $A = 0$, i.e. if the rays from the axial object point are incident normally on the refracting surface, so that object and image coincide at the centre of curvature. This last property is made use of in aplanatic components as in Fig. 8.11, where the refraction at the first surface is aplanatic according to eqn (8.92) and the second surface has normal incidence. Such components are useful in conjunction with that shown in Fig. 8.10 in micro-scope objectives; however, a limitation is that there is always one virtual conjugate, so that a non-aplanatic component is needed if a system with a real object and image is required.

The condition $\Delta(u/n) = 0$ for a mirror reduces to $l = l' = 0$, by eqn (8.92) and recalling from Section 2.2 our convention about signs in reflection: thus this case reduces to the object at the surface, or $h = 0$. The other case, $A = 0$, corresponds to object and image at the centre of curvature and this has many applications.

This last is a particular example of a general rule, that if an optical system

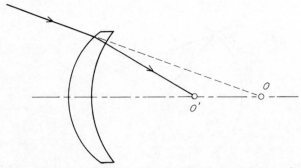

FIG. 8.11. Aplanatic component with normal incidence at the second surface

has complete symmetry about a central plane perpendicular to the axis (which will therefore be the stop plane) and if the magnification is -1 then all aberrations depending on odd powers of the field vanish identically, i.e. coma, distortion and transverse chromatic aberration (Chapter 10) of all orders are zero. This result can be proved by splitting the system into two mirror image parts at the aperture stop and considering the aberration contributions of each part separately with due regard to signs.

There are well-known examples of non-spherical surfaces which are free from spherical aberration. The paraboloid of revolution with one infinite conjugate and the ellipsoid of revolution with conjugates at the foci of the generating ellipse have no spherical aberration, but they do have large coma so they are not suitable for imagery of extended objects. Likewise the hyperboloid of revolution has no spherical aberration for conjugates at the geometrical foci, one of these being a virtual conjugate. A less well-known property of these surfaces is that if the principal ray of an oblique pencil passes through one of the geometrical foci there is no astigmatism for conjugates at any position along the principal ray. The proof of this result depends on the formulae for astigmatism round a finite principal ray given in Chapter 9, Section 9.8; it will be given in that chapter.

9 Finite Aberration Formulae

9.1 Introduction

By a finite aberration formula we mean first an expression for the whole of the aberration for particular aperture and field coordinates, as opposed to the primary aberration expressions with which we have dealt in the previous chapter. Thus for example eqn (7.25), which gives the total wavefront aberration along a certain ray with respect to a reference sphere centred on the intersection of the principal ray with the Gaussian image plane, is a finite aberration expression; likewise eqn (7.29) which was derived from eqn (7.25). Secondly we can extend this definition to include expressions which give the total aberration of a certain kind or kinds; examples of these, which will be derived in this chapter, are the astigmatism formulae of Thomas Young, which give the astigmatism along a finite principal ray, and several expressions akin to the sine condition of Ernst Abbe; these latter deal with the total of coma terms which depend linearly on the field variable.

Finite aberration expressions, of which there is a considerable variety, were developed for one or more of three different purposes: (a) to show the contributions to the aberration from each surface; (b) to obtain, from considerations of symmetry, information about extra-axial aberrations from raytraces of axial pencils; (c) to facilitate computation by reducing the number of significant figures to be carried, just as the primary aberration formulae do. The last point can still be important when computers are available, as noted in Section 7.5, and it suggests a criterion for comparison of different expressions for the same aberration. Suppose that the terms in a certain expression are of degree N in the aperture (and, if applicable, field) coordinates;

then we shall say that we have an Nth-order finite aberration expression. According to this eqn (7.29) is a second order expression and eqn (7.25), since it contains actual distances in the axial direction, is a zero order expression. Clearly no finite aberration expression can be of higher order than 4 if it is for wavefront aberration or 3 if it is for transverse ray aberration, since the primary aberrations are themselves of these orders.

The earliest expressions of this kind seem to have been for ray aberrations of axial pencils and they were given by A. Kerber in 1895. Several years later A. E. Conrady gave similar expressions for wavefront aberration of an axial pencil. These and other formulae are collected with references in a paper by the present author.† We do not reproduce them here because they are arranged for logarithmic–trigonometric calculation and they are therefore of little value for present-day computing methods. We give several more recent results which are more appropriate for computer use, for both ray and wavefront aberrations.

9.2   The Aldis theorem for transverse ray aberrations

One of the most remarkable examples of total aberration expressions is the so-called Aldis theorem, which gives expressions for the transverse ray aberration components $\delta\xi_k'$ and $\delta\eta_k'$ of a finite ray with respect to a paraxial principal ray. As we shall define it the theorem relates to a system of k spherical surfaces. Let $H = n_j u_j \eta_j$ be the Lagrange invariant in the medium preceding the jth surface, for the chosen paraxial field and aperture; let (L_j, M_j, N_j) be the finite ray direction cosines and let (x_j, y_j, z_j) be the co-ordinates of the point of incidence on surface j of this ray. Finally let A_j be the refraction invariant for the paraxial ray from the axial object point, i.e. the ray to which the value of the Lagrange invariant applies. The Aldis theorem states

$$-n_k' u_k' N_k' \, \delta\xi_k' = \sum_{j=1}^{k} \left\{ A_j z_j \Delta L_j + \frac{A_j x_j}{N_j' + N_j} \Delta(L_j^2 + M_j^2) \right\} \tag{9.1}$$

$$-n_k' u_k' N_k' \, \delta\eta_k' = \sum_{j=1}^{k} \left\{ A_j z_j \Delta M_j + \frac{A_j y_j - H}{N_j' + N_j} \Delta(L_j^2 + M_j^2) \right\}. \tag{9.2}$$

The Aldis theorem was given, in a slightly different form, by A. Cox,‡ and ascribed by him to H. L. Aldis. However, it was apparently never published by Aldis. It can easily be seen that eqns (9.1) and (9.2) are third order expressions, according to the classification in Section 9.1. To prove these formulae we note first that in any space we have

$$\eta = y + \frac{M}{N}(l - z), \tag{9.3}$$

† W. T. Welford (1972). *Optica Acta* **19**, 719–727.
‡ A. Cox (1964). "A System of Optical Design", Focal Press, London.

where l is the Gaussian image distance and (ξ, η) is the intersection of the finite ray with the Gaussian image plane. Multiplying through by nuN we have

$$nuN\eta = (A - nhc)Ny - (A - nhc)Mz + nuMl. \tag{9.4}$$

Also we have

$$nuN\eta_0 = HN \tag{9.5}$$

where η_0 refers to the paraxial principal ray involved in the Lagrange invariant H. Then subtracting eqn (9.5) from eqn (9.4) we have

$$nuN\,\delta\eta = A(yN - zM) - nhc(yN - zM) - nhM - HN. \tag{9.6}$$

Thus the increment on refraction of the quantity on the left-hand side is

$$\begin{aligned}\Delta nuN\,\delta\eta &= (Ay - H)\,\Delta N - Az\,\Delta M \\ &\quad - hcy\,\Delta nN + (hcz - h)\,\Delta nM. \end{aligned} \tag{9.7}$$

Now we recall from the finite raytrace equations for spherical surfaces (Section 4.4) that

$$\begin{aligned}\Delta nM &= -cy(n'\cos I' - n\cos I), \\ \Delta nN &= (1 - cz)(n'\cos I' - n\cos I),\end{aligned}$$

so that the third and fourth terms on the right-hand side of eqn (9.7) vanish identically. Thus on summing through the system we obtain

$$n'_k u'_k N'_k\,\delta\eta'_k - n_1 u_1 N_1\,\delta\eta_1 = \sum_{j=1}^{k} \{(A_j y_j - H)\Delta N_j - A_j z_j\Delta M_j\}. \tag{9.8}$$

It was in essentially this form that Cox quoted the result, with the corresponding equation for $\delta\xi'_k$; it can be seen that on account of the expression ΔN this is a first order formula. However, if we apply the method of transformation given in Section 7.5 (eqn (7.28)), we can obtain

$$\Delta N = -\frac{\Delta(L^2 + M^2)}{N' + N}$$

and we obtain eqn (9.2) above, which is a third order formula. The derivation of eqn (9.1) follows exactly the same lines. In eqn (9.8) we have included on the left the term $n_1 u_1 N_1\delta\eta_1$ which occurs formally in the summation; this would usually be zero, but it can be regarded as representing the aberration in an incoming pencil from a preceding optical system.

Cox also gave a simple extension of the Aldis theorem to include aspheric surfaces. Let the equation of an aspheric be given in the form

$$F(x, y, z) \equiv 2z - c(x^2 + y^2 + z^2) + f(x^2 + y^2) = 0; \tag{9.9}$$

here c is the paraxial curvature and f is a term representing the aspherizing.

The direction cosines of the normal are then

$$(\alpha, \beta, \gamma) = \frac{2(-x(c - f'), -y(c - f'), 1 - zc)}{\{F_x^2 + F_y^2 + F_z^2\}^{\frac{1}{2}}} \tag{9.10}$$

and we have from Chapter 4 that

$$\Delta n(L, M, N) = (\alpha, \beta, \gamma)\Delta n \cos I.$$

If this result is used in eqn (9.7) we find that the terms corresponding to the spherical part of the equation of the surface again vanish and we have finally for the two components of aberration the expressions given in eqns (9.1) and (9.2) with the following additional terms for the aspherizing at each aspheric surface:

$$\left.\begin{array}{ll} \dfrac{2hx(cz - 1)f'}{\{F_x^2 + F_y^2 + F_z^2\}^{\frac{1}{2}}} \Delta n \cos I & (x \text{ component}) \\[4mm] \dfrac{2hy(cz - 1)f'}{\{F_x^2 + F_y^2 + F_z^2\}^{\frac{1}{2}}} \Delta n \cos I & (y \text{ component}) \end{array}\right\}. \tag{9.11}$$

Since the aspherizing polynomial must be at least of order 2 in $x^2 + y^2$ these contribute 3rd order terms to the transverse ray aberrations.

The terms in the Aldis theorem summation are dimensionally lengths. It is possible to paraxialize these expressions and so arrive at an alternative derivation, one of many, for the Seidel aberrations; this would, of course, lead to transverse ray aberration expressions but the summations involved are essentially the same as for wavefront aberrations in the Seidel domain so that the Seidel sums would be obtained.

9.3 Expressions for total optical path aberration

Historically, expressions for total optical path aberration were developed first for axial pencils, then for off-axis pencils in a meridian plane and finally for skew rays. The derivation for skew rays is actually quite straightforward and it is easy to derive the more specialized cases from the skew ray formula; we therefore derive the expression for skew rays first.

Let r (Fig. 9.1) be a skew ray of an off-axis pencil and let $\bar{r}$ be the (finite) principal ray of this pencil; let r and $\bar{r}$ meet the refracting surface in P and $\bar{P}$ and the Gaussian image plane in Q and $\bar{Q}$ and let the perpendicular from $\bar{Q}$ to the ray r meet it in T. Let a and $\bar{a}$ be position vectors from the vertex of the refracting surface to P and $\bar{P}$ and let r and $\bar{r}$ also denote, as usual, unit vectors along the respective rays. The expression

$$\Delta n(\bar{P}\bar{Q} - PT) \tag{9.12}$$

gives the difference in optical paths along the two rays measured respectively from $\bar{Q}$ to $\bar{Q}'$ and from T to T'; that is to say, it gives the wavefront aberration for the skew ray path for foci, i.e. centres of reference spheres, at $\bar{Q}$ and $\bar{Q}'$ provided the reference spheres are regarded as having infinite radii. (Alternatively this expression is the eikonal referred to $\bar{Q}$ and $\bar{Q}'$ as origins; however, it is usual in formal aberration theory to refer the eikonal only to origins

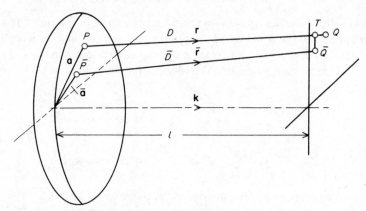

FIG. 9.1. Notation for total optical path aberration

on the optical axis.) Thus the expression (9.12) if summed over all surfaces is a total aberration formula for optical path aberration but, of course, it is only a zero order formula.

We can convert (9.12) to a second order expression: let $\bar{P}\bar{Q} = \bar{D}$ and $PQ = D$; the vector displacement $\bar{Q}Q$ is easily seen to be equal to $a + Dr - \bar{a} - \bar{D}\bar{r}$, so that

$$\bar{P}\bar{Q} - PT = \bar{D} - D + (a + Dr - \bar{a} - \bar{D}\bar{r})\cdot r$$
$$= \bar{D}(1 - \bar{r}\cdot r) + r\cdot(a - \bar{a}). \tag{9.13}$$

But

$$\bar{D} = \frac{l - \bar{a}\cdot k}{\bar{r}\cdot k}, \tag{9.14}$$

where l is the Gaussian image distance and k is a unit vector along the optical axis, and also

$$1 - \bar{r}\cdot r = \frac{1 - (\bar{r}\cdot r)^2}{1 + \bar{r}\cdot r}$$

$$= \frac{|\bar{r} \wedge r|^2}{1 + \bar{r}\cdot r}. \tag{9.15}$$

If we substitute from eqns (9.14) and (9.15) into eqn (9.13) we have

$$\Delta n(\overline{PQ} - PT) = \Delta n \left\{ \frac{(l - \bar{a} \cdot k)|\bar{r} \wedge r|^2}{(1 + \bar{r} \cdot r)\bar{r} \cdot k} + r \cdot (a - \bar{a}) \right\}. \qquad (9.16)$$

Since $|\bar{r} \wedge r|^2$ is the square of the sine of the angle between the two rays it is a second order quantity; also $r \cdot a$ is second order and so (9.16) when summed over all surfaces gives a second order aberration formula for skew ray optical paths, referred to the intersection of the principal ray with the Gaussian image plane. In terms of the more familiar cartesian components used, for example, in Chapter 4, eqn (9.16) becomes

$$\Delta n(\overline{PQ} - PT) = \Delta n \left\{ (l - \bar{z}) \frac{\{(M\bar{N} - \bar{M}N)^2 + (N\bar{L} - \bar{N}L)^2 + (L\bar{M} - \bar{L}M)^2\}}{\bar{N}(1 + L\bar{L} + M\bar{M} + N\bar{N})} \right.$$

$$\left. + L(x - \bar{x}) + M(y - \bar{y}) + N(z - \bar{z}) \right\}. \qquad (9.17)$$

So far no method of transforming this into a higher order expression has been found. It is necessary to adopt a different approach to get a fourth order expression suitable for skew rays. We recall first that the Aldis theorem gives transverse ray aberration components with respect to the Gaussian image point, a constant reference point for all rays; likewise eqn (9.16) gives the optical path differences for all finite rays with respect to a constant reference point, the intersection of the finite principal ray with the Gaussian image plane. However, a common characteristic of all fourth order total aberration expressions for optical path differences turns out to be that the reference point or centre of the reference sphere is defined by the particular pair of rays between which the optical path difference is being found and thus it is not a constant point for a given pencil. Thus for an axial pencil the reference point is the intersection of the finite ray and the axis and for two rays in a meridian plane it is the intersection point of the two rays. Thus if several rays have been traced and the optical path differences with respect to a common principal ray are found a focal shift operation has to be carried out for each ray to get to a common focal point for the whole pencil. For a skew ray and a principal ray the question arises, what point should be chosen as the reference point, since there is no intersection? H. H. Hopkins suggested the mid-point of the common perpendicular, or the shortest join, of the two rays and this led to a very compact second order formula.† Let two rays r and $\bar{r}$ meet the refracting surface at points with position vectors a and $\bar{a}$ (Fig. 9.2); the ray $\bar{r}$ may be thought of as the principal ray but in fact the result to be proved holds if both are skew rays. A general point on the ray r is $a + Dr$ where D is the distance

† H. H. Hopkins (1952). *Proc. phys. Soc.* **LXVB**, 934–942.

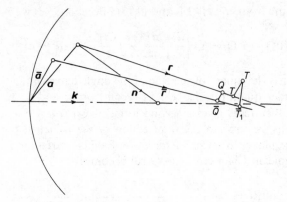

FIG. 9.2. Optical path difference to the common perpendicular of two rays

of this point from the point of incidence, and similarly for the other ray. The distance s between these two points is given by

$$s^2 = |a + Dr - \bar{a} - \bar{D}\bar{r}|^2. \tag{9.18}$$

The values of D and $\bar{D}$ corresponding to Q and $\bar{Q}$, the feet of the common perpendicular to the two rays, are obtained by differentiating s^2 with respect to D and $\bar{D}$ and setting the partial derivatives equal to zero. We obtain

$$\left.\begin{aligned} D &= \frac{-(r - (r\cdot\bar{r})\bar{r})\cdot(a - \bar{a})}{1 - (r\cdot\bar{r})^2} \\[2mm] \bar{D} &= \frac{(\bar{r} - (r\cdot\bar{r})r)\cdot(a - \bar{a})}{1 - (r\cdot\bar{r})^2} \end{aligned}\right\} \tag{9.19}$$

and from these we have

$$\bar{D} - D = \frac{(r + \bar{r})\cdot(a - \bar{a})}{1 + r\cdot\bar{r}}. \tag{9.20}$$

We can now take $[\bar{Q}\bar{Q}'] - [QQ']$ as a measure of the optical path difference introduced on refraction through the current surface; this is, from eqn (9.20),

$$\Delta W = \Delta n \frac{(r + \bar{r})\cdot(a - \bar{a})}{1 + r\cdot\bar{r}} \tag{9.21}$$

and the total path difference between the rays is obtained by summing this expression through the system. Equation (9.21) is Hopkins' result; he showed that it gave the increment of wavefront aberration referred to reference spheres centred at the midpoints of $\bar{Q}Q$ and $\bar{Q}'Q'$ but in the present context it seems more appropriate to regard eqn (9.21) as expressing the difference in optical paths along the two rays measured to the common perpendicular, since the exact summation through the system can then be seen immediately.

The expression on the right-hand side of eqn (9.21) is still only a second order expression, since the denominator is zero order and the terms such as $r \cdot a$ in the numerator contain Lx and My, products of first order quantities, and Nz, product of a zero and a second order quantity. It would reduce identically to zero if paraxialized. It is a reasonable conjecture that it is necessary to use Snell's law to obtain a fourth order expression for optical path differences; certainly this is so for all known fourth order expressions and we did not make use of Snell's law in deriving eqn (9.21).

We now transform (9.21) to 4th order form. We have first

$$\frac{1}{1 + r \cdot \bar{r}} = \frac{1}{2}\left(1 + \frac{1 - r \cdot \bar{r}}{1 + \bar{r} \cdot r}\right)$$

$$= \frac{1}{2}\left(1 + \frac{(1 - (r \cdot \bar{r})^2)}{(1 + \bar{r} \cdot r)^2}\right)$$

$$= \frac{1}{2}\left(1 + \frac{|r \wedge \bar{r}|^2}{(1 + r \cdot \bar{r})^2}\right).$$

Substituting in eqn (9.21) we have

$$\Delta W = \tfrac{1}{2}\Delta n(r + \bar{r}) \cdot (a - \bar{a})\left\{1 + \frac{|r \wedge \bar{r}|^2}{(1 + r \cdot \bar{r})^2}\right\}. \tag{9.22}$$

We now consider only the quantity $\Delta W_1 = \tfrac{1}{2}\Delta n(r + \bar{r}) \cdot (a - \bar{a})$, since the rest of the expression in (9.22) is of 4th order. Let n and $\bar{n}$ be unit vectors along the normals at the points of incidence in Fig. 9.2 and let k be a unit vector along the optical axis. We have $a = r(k - n)$ where r is the radius of curvature of the refracting surface; thus on substituting for a and $\bar{a}$,

$$\Delta W_1 = \tfrac{1}{2}r\Delta n(r + \bar{r}) \cdot (\bar{n} - n)$$
$$= \tfrac{1}{2}r(\bar{n} - n) \cdot \Delta n(r + \bar{r}). \tag{9.23}$$

Now Snell's law in vector form gives (eqn (4.1))

$$\Delta nr = n\Delta n(r \cdot n) \tag{9.24}$$

and substituting for Δnr and $\Delta n\bar{r}$ in eqn (9.23) we have

$$\Delta W_1 = \tfrac{1}{2}r(\bar{n} - n) \cdot \{n\Delta n(r \cdot n) + \bar{n}\Delta n(\bar{r} \cdot \bar{n})\}$$
$$= -\tfrac{1}{2}r(1 - n \cdot \bar{n})\Delta n(r \cdot n - \bar{r} \cdot \bar{n})$$
$$= -\tfrac{1}{2}r\frac{|n \wedge \bar{n}|^2}{1 + n \cdot \bar{n}}\Delta n\frac{\{|\bar{r} \wedge \bar{n}|^2 - |r \wedge n|^2\}}{\bar{r} \cdot \bar{n} + r \cdot n}. \tag{9.25}$$

Since the denominators are of zero order this is a fourth order expression; it is, however, slightly unsatisfactory still because the left-hand factor, $r|\mathbf{n} \wedge \bar{\mathbf{n}}|^2/(1 + \mathbf{n}\cdot\bar{\mathbf{n}})$ becomes indeterminate, in the form $\infty \times 0$, as the radius r tends to infinity. To put this right we recall that this term was obtained in the last step in eqn (9.25) from the expression $r(1 - \mathbf{n}\cdot\bar{\mathbf{n}})$; if we substitute for the normals from $\mathbf{a} = r(\mathbf{k} - \mathbf{n})$ we obtain

$$\frac{r|\mathbf{n} \wedge \bar{\mathbf{n}}|^2}{1 + \mathbf{n}\cdot\bar{\mathbf{n}}} = r(1 - \mathbf{n}\cdot\bar{\mathbf{n}})$$

$$= \mathbf{k}\cdot(\mathbf{a} + \bar{\mathbf{a}}) - \frac{\mathbf{a}\cdot\bar{\mathbf{a}}}{r}, \tag{9.26}$$

which is well behaved for all r. We can now use this in eqn (9.25) and bring back the 4th order term from (9.22) which was temporarily discarded, to obtain the final result:

$$\Delta W = -\frac{1}{2}\left\{\mathbf{k}\cdot(\mathbf{a} + \bar{\mathbf{a}}) - \frac{\mathbf{a}\cdot\bar{\mathbf{a}}}{r}\right\}\Delta n\frac{(|\bar{\mathbf{r}} \wedge \bar{\mathbf{n}})|^2 - |\mathbf{r} \wedge \mathbf{n}|^2)}{\bar{\mathbf{r}}\cdot\bar{\mathbf{n}} + \mathbf{r}\cdot\mathbf{n}}$$

$$+ \tfrac{1}{2}\Delta n(\mathbf{r} + \bar{\mathbf{r}})\cdot(\mathbf{a} - \bar{\mathbf{a}})\frac{|\mathbf{r} \wedge \bar{\mathbf{r}}|^2}{(1 + \mathbf{r}\cdot\bar{\mathbf{r}})^2} \tag{9.27}$$

This, a true fourth order result, was given by the present author (reference in Section 9.1). On summing through the system it gives the total optical path difference from the object point to the ends of the shortest join of the two rays in image space; alternatively this is the wavefront aberration with respect to a reference sphere centred on the midpoint of this shortest join. Equation (9.27) can be paraxialized to give the Seidel sums; this is an involved operation since the expression obtained at first contains the quantities $\bar{u}$ and $\bar{h}$ and it is necessary to replace these by expressions involving $\bar{A}$, the refraction invariant of the principal ray, and H, the Lagrange invariant; the appropriate substitutions are given in Chapter 6.

Equation (9.27) is, of course, valid also for meridian rays. Thus if we put $\bar{\mathbf{a}} = 0, \bar{\mathbf{r}} = \bar{\mathbf{n}} = \mathbf{k}$, we obtain an expression for the path difference along a ray of an axial pencil:

$$\Delta W_{\text{axial}} = \tfrac{1}{2}z\Delta\frac{n \sin^2 I}{1 + \cos I}$$

$$+ \tfrac{1}{2}\Delta n\frac{(My + (1 + N)z)}{(1 + \cos U)^2}\sin^2 U. \tag{9.28}$$

Here the vector terms have been replaced by their scalar equivalents, the coordinates, convergence angles, etc. of the finite meridian ray. Expressions of this kind but adapted specifically for logarithmic computation are collected in the reference in Section 9.1.

As noted above we might wish to refer the aberration as obtained from eqn (9.27) or (9.28) in image space to another image point, e.g. the final Gaussian image point. Then a term for longitudinal and transverse focal shift has to be added to the summation of terms like eqn (9.27). Focal shifts are given by second order expressions and since the shift concerned can be quite arbitrary, i.e. not particularly related to the pair of rays to which eqn (9.27) refers, this final term cannot be made higher than second order. Nevertheless the overall saving in numerical precision from using a fourth order expression for the main summation is considerable.

To get an expression suitable for use with eqn (9.27) let the new focus be T in Fig. 9.2 and let its position vector be p, referred to the same origin as $\bar{a}$ and a. Let $\bar{T}_1$ and T_1 be the projections of T on the rays $\bar{r}$ and r. Then assuming as before a reference sphere of infinite radius the required increment of aberration is

$$\delta W = n'(\bar{Q}\bar{T}_1 - QT_1) \qquad (9.29)$$

where it is assumed also that we are in the final image space and that n' is the refractive index in that space. But we have

$$\bar{Q}\bar{T}_1 = (p - (\bar{a} + \bar{P}\bar{Q}\bar{r})) \cdot \bar{r}$$

$$QT_1 = (p - (a + PQr)) \cdot r$$

so that

$$\delta W/n' = p \cdot (\bar{r} - r) + (\bar{a} \cdot \bar{r} - a \cdot r) - (\bar{P}\bar{Q} - PQ). \qquad (9.30)$$

The second term is clearly of second order; the third term is given by eqn (9.20) and is also second order. The first term can be seen to be second order by writing it in components as follows.

$$p \cdot (\bar{r} - r) = p_x(\bar{L} - L) + p_y(\bar{M} - M) + p_z(\bar{N} - N)$$

$$= p_x(\bar{L} - L) + p_y(\bar{M} - M) + \frac{p_z(L^2 + M^2 - \bar{L}^2 - \bar{M}^2)}{\bar{N} + N}. \qquad (9.31)$$

This completes the expression of an arbitrary focal shift, transverse and longitudinal, in second order terms.

9.4 Aplanatism and isoplanatism

The general concept of isoplanatism plays an important role in the theory of the formation of images of extended objects, both according to physical theory and in the geometrical optics approximation. An optical system is said to be isoplanatic if a displacement of the object point produces no change in the aberrations of the corresponding pencil in image space, referred, of course, to the appropriately shifted ideal image point. This concept would normally apply to a limited region of the field and usually only a differential, i.e. an infinitesimally small, object point displacement is considered. Aplanatism is the corresponding stationarity condition when the pencil has actually zero aberration at the original object point. The significance of the concept of iso-planatism in the theory of image formation of extended objects is that if the form of the point image (the so-called point spread function) is known for one point of the object it is known everywhere, if the system is isoplanatic over the relevant region of the field; then the image of the complete object can be found by a relatively simple convolution integral, rather than having to allow for a varying spread function.

The term *aplanatism* was used by early nineteenth century writers, e.g. Coddington and Herschel, to mean simply freedom from spherical aberration. The present day usage in the sense of stationarity of aberration as well as freedom from spherical aberration was begun by Ernst Abbe. Abbe's first paper concerned with aplanatism, among other things, appeared in 1873, in Schultze's *Archiv für mikroskopische Anatomie* **IX**, 413–468. In this he gave the *sine condition*, which is the condition for axial aplanatism. The term iso-planatism appeared in the literature in 1919 when F. Staeble and E. Lihotzky simultaneously published conditions for zero linear coma in systems with spherical aberration.

Axial symmetry in an optical system makes it possible to estimate departure from isoplanatism near the axis by very simple formulae involving only rays of the axial pencil; thus the concept of isoplanatism is important in optical design. In addition to these simpler formulae we shall also give some more general results relating to isoplanatism in systems with no symmetry; these results have, apart from their intrinsic theoretical interest, applications to the geometrical optics of holographic image formation.

9.5   Linear coma and offence against the sine condition

After Abbe published his sine condition in 1873 several other authors gave different proofs, and then came variations and generalizations of the basic theorem.

We shall give here a derivation of the most useful of these and show its relationship to the other results. Abbe's result and all the others relate to *linear coma*, i.e. coma varying linearly with the field coordinate. If we refer to Section 7.7 we see that terms of the form $y\eta'(x^2 + y^2)^m$ occur in the aberration expansion; these are all like Seidel coma, for which $m = 1$, in that they show the same degree of symmetry, the same azimuthal variation in the pupil and the same, linear, dependence on field. The sum of terms of this type

$$W_{\text{coma}} = \sum_m c_m(x^2 + y^2)^m \eta' y \tag{9.32}$$

is therefore called the linear coma and it is this which is to be evaluated. We suppose, as usual, that x and y are exit pupil coordinates and we assume a circular pupil of radius a'. The y-component of transverse ray aberration due to linear coma is (eqn (7.15)).

$$-\frac{R'\eta'}{n'} \sum_m c_m(x^2 + (2m + 1)y^2)(x^2 + y^2)^{m-1}, \tag{9.33}$$

where R' is the radius of the reference sphere, and the value of this for the point $(a', 0)$ in the pupil is

$$\delta\eta'_{\text{sag.}} = -\frac{R'\eta'}{n'} \sum_m c_m a'^{2m}$$

$$= -\frac{R'}{n'a'} W_{\text{coma}}(0, a') \tag{9.34}$$

from eqn (9.32). This above result can be used to compute W_{coma} from the axial raytrace because $\delta\eta'_{\text{sag.}}$ can be found as follows. Let P be the axial point of the Gaussian image plane (Fig. 9.3) and let P_1 be the axial focus of the rays

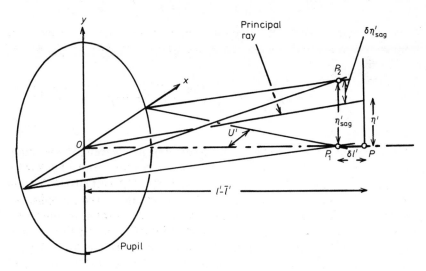

FIG. 9.3. Total linear coma calculated from the axial pencil

from the rim of the pupil; thus P_1P is the longitudinal spherical aberration $\delta l'$, taken positive as in the diagram, according to accepted convention. With our usual notation we have $R' = l' - \bar{l}'$ and then from the diagram

$$\delta\eta'_{\text{sag.}} = \eta'_{\text{sag.}} - \left(\frac{l' - \bar{l}' - \delta l'}{l' - \bar{l}'}\right)\eta', \tag{9.35}$$

where η' is the Gaussian image height and $\eta'_{\text{sag.}}$ is the height above the axis at which the sagittal pair of rays intersect in P_2. But from the optical sine theorem (eqn (6.31)) we have

$$n\eta \sin U = n'\eta'_{\text{sag.}} \sin U', \tag{9.36}$$

where U and U' are convergence angles of the axial pencil at full aperture. Thus from eqns (9.34, 9.35, 9.36) we have

$$W_{\text{coma}}(0, a') = -\frac{n'a'}{R'}\left\{\frac{n\eta \sin U}{n' \sin U'} - \eta'\frac{(l' - \bar{l}' - \delta l')}{l' - \bar{l}'}\right\},$$

or, since $a'/R' = -\sin U'$,

$$W_{\text{coma}}(0, a') = n\eta \sin U - n'\eta' \sin U'\left(\frac{l' - \bar{l}' - \delta l'}{l' - \bar{l}'}\right). \tag{9.37}$$

If we take out the Lagrange invariant $H = nu\eta$ as a factor on the right-hand side we obtain

$$W_{\text{coma}}(0, a') = -H\left\{\frac{\sin U'}{u'}\left(\frac{l' - \bar{l}' - \delta l'}{l' - \bar{l}'}\right) - \frac{\sin U}{u}\right\}. \tag{9.38}$$

Since as noted above all the components of linear coma have the same azimuthal variation in the pupil, namely as $\cos \phi$, where $\tan \phi = x/y$, we can write finally

$$W_{\text{coma}}(a' \sin \phi, a' \cos \phi) = -H \cos \phi\left\{\frac{\sin U'}{u'}\left(\frac{l' - \bar{l}' - \delta l'}{l' - \bar{l}'}\right) - \frac{\sin U}{u}\right\}. \tag{9.39}$$

This remarkable expression gives the linear coma as a wavefront aberration in terms of the properties of the axial pencil alone, together with a knowledge of the position of the pupil. From our formulae for stop-shift effects (Section 8.5) we expect coma to be dependent on the stop position if there is spherical aberration; conversely if $\delta l'$ is zero eqn (9.39) shows that the linear coma does not depend on the pupil position. Equation (9.39) refers strictly to the coma at the radius in the pupil for which the finite ray of the axial pencil is traced; other rays will behave differently according to the combinations of higher order spherical aberration and coma present in the system. Also eqn (9.39) refers only to the region near enough to the axis to justify neglecting effects depending on higher powers than the first in η. Non-linear coma-like aberrations could be of the form $(x^2 + y^2)^m \eta^{2n+1} y$ where n is non-zero, and these are not taken account of by eqn (9.39).

If the object plane tends to infinity, it can easily be seen that the second term in the brackets on the right-hand side of eqn (9.39) tends to y/h where y is the finite ray height in the entrance pupil and h is the paraxial ray height.

The ratio $\delta\eta'_{sag.}$ ($\eta'_{sag.} - \delta\eta'_{sag.}$) is a useful dimensionless measure of linear coma; A. E. Conrady ("Applied Optics and Optical Design") called it the offence against the sine condition (OSC') and gave the following expression for it in 1929:

$$\text{OSC}' = \frac{\sin U}{u} \cdot \frac{u'}{\sin U'} \left(\frac{l' - \bar{l}'}{l' - \bar{l}' - \delta l'} \right) - 1. \tag{9.40}$$

When, as is usually the case, $u = \sin U$ this takes the simple form

$$\text{OSC}' = \frac{u'}{\sin U'} \left(\frac{l' - \bar{l}'}{l' - \bar{l}' - \delta l'} \right) - 1. \tag{9.41}$$

The above results are expressions for aberrations and they thus give directly magnitudes which may be tested against tolerance figures. Microscope objectives, telescope objectives and collimators are examples of systems with small fields where all but the final stages of design can be carried out using such expressions to control the off-axis aberrations. A different view is obtained by re-casting the expressions as conditions for zero coma. We then obtain from eqn (9.39)

$$\frac{\sin U'}{u'} \left(\frac{l' - \bar{l}' - \delta\bar{l}'}{l' - \bar{l}'} \right) = \frac{\sin U}{u} \tag{9.42}$$

as the condition for isoplanatism for the zone of the pupil at which the finite ray was traced. If we assume no spherical aberration eqn (9.42) reduces to the form

$$\frac{\sin U'}{u'} = \frac{\sin U}{u}; \tag{9.43}$$

this is the original sine condition of Abbe, the condition for aplanatism of a symmetrical optical system. As before, it applies strictly to the aperture at which the finite ray with convergence angles U and U' is traced; there may be longitudinal spherical aberration at other apertures and even if not, eqn (9.43) may not be satisfied. Isoplanatism and aplanatism conditions of this kind have applications in the synthesis of optical systems, i.e. design from closed formulae rather than by raytracing and successive approximations. For example, there is a class of two-surface systems, i.e. two mirrors, a single thick lens or two separated thin lenses, for which analytic solutions for exact aplanatism exist if both elements are aspheric. Equation (9.42) can also be used to solve for the position of the stop which will give zero coma, although the cases in which this is of practical use are rather rare.

Equation (9.42) (or various rearrangements of it) is also known as the Staeble–Lihotzky condition; it is given with variants by Berek,[†] who discusses the variations and their significances in detail.

9.6   Isoplanatism in non-symmetric systems

The sine condition and its several generalizations discussed above are concerned with the region near the axis of a symmetrical optical system. We may also be interested in stationarity of aberrations at a finite distance from the axis, i.e. we may ask under what conditions the shape of an off-axis pencil is unchanged for a small change in the (finite) field angle. This leads to a more general question in which no symmetry at all is assumed in the system and we consider the effect on an emergent pencil of a small displacement of the incident pencil. Thus, referring to Fig. 9.4, the wavefront Σ incident on an arbitrary

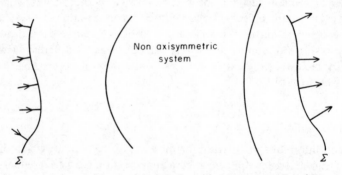

FIG. 9.4. Isoplanatism in an optical system with no symmetry; Σ and Σ' are incident and emergent wavefronts belonging to a single pencil

optical system, restricted only to having isotropic media, emerges as Σ'; under what conditions will a small rigid displacement of Σ result in a rigid displacement of Σ' without a change in shape? Some attempts at solutions of this problem under various restricted conditions have been made but it was not until 1971 that the problem was stated and solved in complete generality, by the present author (*Optics Communications*, **3**, 1–6). It has considerable

---

† M. Berek (1930). "Grundlagen der praktischen Optik", de Gruyter, Leipzig.

importance in the theory of holographic image formation and several special cases can be deduced which are relevant to the symmetrical optical system.

The most general possible displacement of the wavefront Σ in Fig. 9.4 (together with the associated rays and the other wavefronts of the *pencil*) is an infinitesimal rotation about an arbitrarily chosen axis; if the system is isoplanatic with respect to this axis then such an infinitesimal rotation will produce an infinitesimal rotation of the emergent wavefront Σ' about some axis in image space *without change of shape of* Σ'. In order to obtain the required condition, which in fact turns out to be remarkably simple, we need first an auxiliary theorem on infinitesimal rotations of wavefronts.

Let $V(a_0, a)$ be the characteristic function (Section 7.2) of the part of the optical system between the object space and some intermediate space; we take a_0 as the position vector of an object point P_0 and a as the position vector of any point in the intermediate space. The equation of a wavefront Σ in this intermediate space is

$$V(a_0, a) = C, \tag{9.44}$$

where C is the optical path length from P_0 to the wavefront and the coordinates (x, y, z) in this implicit equation are, of course, the components of a. Let p be a unit vector along a proposed axis of rotation in the space in which the wavefront Σ lies (Fig. 9.5) and suppose the wavefront (and the rest of the pencil) to be rotated through a small angle ε about p to a new position Σ_1; let b be the position vector of any point on the axis p.† To find the position of Σ_1 we note that a point with position vector a_1 on Σ_1 is obtained from the point with position vector a on Σ by adding the vector $\varepsilon p \wedge (a - b)$, neglecting squares of ε. Thus the equation of Σ can also be written

$$V(a_0, a_1 - \varepsilon p \wedge (a - b)) = C + O(\sigma^2), \tag{9.45}$$

where σ is a quantity of order of magnitude $\varepsilon|a - b|$. Now expanding the left-hand side by Taylor's theorem as far as the linear term this becomes

$$V(a_0, a_1) - \varepsilon(p \wedge (a - b)) \cdot \nabla V = C + O(\sigma^2), \tag{9.46}$$

where the gradient is with respect to a. But ∇V is simply nr, where r is a unit vector along the ray, by the basic property of the characteristic function (Section 7.2); thus the rotated surface is given by the position vector a_1, where

$$V(a_0, a_1) - \varepsilon n\{p, (a - b), r\} = C + O(\sigma^2); \tag{9.47}$$

---

† The surface Σ_1 so defined is not necessarily a physically possible wavefront for this system, in the sense that there may not be a single object point displaced from P_0 from which all the rays of Σ_1 could have originated; however, since it is a surface of the same shape as Σ there will be a congruence of rays normal to it and these rays when produced back to the object space will be an aberrated pencil which could have been produced by a further optical system of suitable properties. Thus it is reasonable to use a notation which implies that Σ_1 is a wavefront.

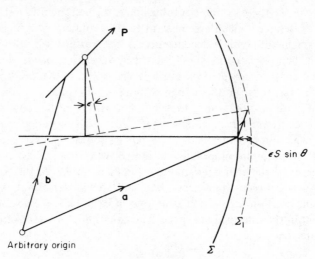

FIG. 9.5. A small rotation of a wavefront; the rotation is through the angle
ε about the axis p

in this equation the brackets { } denote the triple scalar product of the three
vectors; this is equal to the volume of the parallelepiped with sides p, $a - b$
and r and since p and r are unit vectors this volume is equal numerically to
the length S of the common perpendicular to p and r multiplied by the sine
of the angle θ between p and r. Equation (9.47) can thus be written

$$V(a_0, a_1) = C + \varepsilon n S \sin \theta + O(\sigma^2). \tag{9.48}$$

This is the required auxiliary theorem. It means that the rotated wavefront
Σ_1 can be obtained from the original wavefront Σ by marking off along each
ray the distance $\varepsilon S \sin \theta$, as in Fig. 9.5, neglecting quantities of order σ^2. The
sign of ε must, of course, be chosen correctly to represent the direction of
rotation about the axis p. We note that the position vector a of a point on the
wavefront does not appear in eqn (9.48); this is as it should be since we should
not expect the choice of any particular wavefront of the pencil to affect the
result; in other words, we are really concerned with the change in optical
path along a particular ray, that to which the given values of S and $\sin \theta$
refer. We now state the general isoplanatism theorem mentioned above.

Let p and p' be unit vectors along rotation axes in object and image spaces,
let r and r' be unit vectors along segments of a ray of a pencil in object and
image space, let θ and θ' be the angles between r and p and between r' and
p' and let S and S' be the lengths of the common perpendiculars to or
shortest joins between r and p and r' and p'. Then the necessary and suffi-

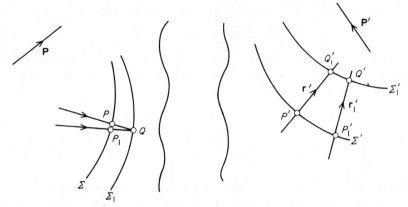

FIG. 9.6. The most general isoplanatism theorem

cient condition that a small rotation ε of the object pencil about p shall produce a small rotation ε' of the image pencil about p' is

$$\varepsilon' n' S' \sin \theta' = \varepsilon n S \sin \theta + A + O(\sigma^2) \tag{9.49}$$

for all rays of the pencil, where A is a constant. Furthermore if the condition is not fulfilled the increment in optical path aberration due to the rotation is

$$\delta W = \varepsilon' n' S' \sin \theta' - \varepsilon n S \sin \theta + O(\sigma^2). \tag{9.50}$$

To prove this theorem, let Σ and Σ' be wavefronts of the pencil, as in Fig. 9.6 and let Σ become Σ_1 after a rotation ε about the axis p. Let Σ_1' be a wavefront in image space of the pencil formed by the normals to the surface Σ_1, i.e. the rays of the pencil, and let $r_1 \ldots r_1'$ be a ray of this new pencil. We choose r_1 so that it meets Σ_1 at the same point Q as r, as in the figure; also r meets Σ, Σ' and Σ_1' in P, P' and Q_1' while r_1 meets these surfaces in P_1, P_1' and Q'. We choose the wavefront Σ_1' to be that one of the pencil from Σ_1 which satisfies the condition

$$[QQ']_{r_1} = [PP']_r. \tag{9.51}$$

Now

$$[QQ']_{r_1} = [P_1 P_1']_{r_1} + [P_1' Q'] - [P_1 Q], \tag{9.52}$$

these distances all being measured algebraically along the ray r_1, and since the angle between the two rays is of order ε in all spaces we have by Fermat's principle

$$[P_1 P_1']_{r_1} = [PP']_r + O(\sigma^2),$$
$$[P_1 Q] = [PQ] + O(\sigma^2),$$
$$[P_1' Q'] = [P' Q_1'] + O(\sigma^2).$$

Combining these with eqns (9.51) and (9.52) we obtain

$$[P'Q_1'] = [PQ] + O(\sigma^2). \tag{9.53}$$

We have from the auxiliary theorem (eqn (9.48)), $[PQ] = \varepsilon n S \sin \theta$ and if Σ_1' is in fact a wavefront of a pencil which is obtained from Σ' by a rigid rotation ε' about p' then, again from eqn (9.48),

$$[P'Q_1'] = \varepsilon' n' S' \sin \theta' + A; \tag{9.54}$$

the constant A occurs here because Σ_1' may be the wavefront obtained by rotating a wavefront parallel to Σ' rather than Σ' itself. Thus the necessity of the theorem is proved.

To prove that the condition is sufficient we note that if eqn (9.49) (the theorem) is true for all rays then $[P'Q_1']$ must, by reversing the above reasoning, be given by an equation of the form of eqn (9.54) above; thus Σ_1' must be parallel to a surface obtained by a rigid rotation of Σ'. This completes the proof that eqn (9.49) is a necessary and sufficient condition for isoplanatism.

To prove the second part of the theorem, suppose eqn (9.49) is not fulfilled; let Σ_1' be, as before, a wavefront corresponding to the rotated incident pencil Σ_1 as in Fig. 9.7. We can rotate Σ' through ε' about the assigned axis p' to give a surface Σ_2' which would have been a wavefront parallel to Σ_1' if the condition had been satisfied. Let the ray r' meet Σ_2' in Q_2', the other points P' and Q_1' being defined as before; then by the auxiliary theorem

$$[P'Q_2'] = \varepsilon' n' S' \sin \theta' + A \tag{9.55}$$

and we have also

$$[P'Q_2'] = [P'Q_1'] - [Q_2'Q_1']. \tag{9.56}$$

Thus from eqn (9.53)

$$[Q_2'Q_1'] = [PQ] - [P'Q_2'] \tag{9.57}$$

$$= \varepsilon' n' S' \sin \theta' - \varepsilon n S \sin \theta + A. \tag{9.58}$$

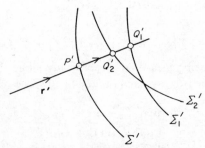

FIG. 9.7. Departure from isoplanatism in a system with no symmetry

Since $[Q'_2 Q'_1]$ gives the change in shape of the wavefront from the purely rotated position Σ'_2 this proves the second part of the theorem.

The constant A in the formulation of the isoplanatism theorem is of no great importance; it occurs because there is a free choice of which wavefront shall be considered. In applying the theorem the constant always disappears, for the condition must always be tried for at least two different rays and it is the difference between these which matters. The theorem applies to any image forming system to which Fermat's principle applies; there are also systems such as diffraction gratings and holograms in which a simple extension of Fermat's principle applies and our theorem applies to these also with appropriate modifications. In considering actual cases it is usual to obtain the quantity $S \sin \theta$ directly as the triple scalar product $\{p, D, r\}$ where D is a displacement vector from any point of p to any point of r; thus in our derivation above $D = a - b$; there is generally an obvious choice for D, as in the examples we shall give.

9.7 Applications of the general isoplanatism theorem of Section 9.6

Consider first an axial pencil in a symmetrical optical system. Taking rectangular axes at the exit pupil, as in Fig. 9.8, let l' be the paraxial image distance from the pupil and $\delta l'$ the longitudinal spherical aberration of a finite

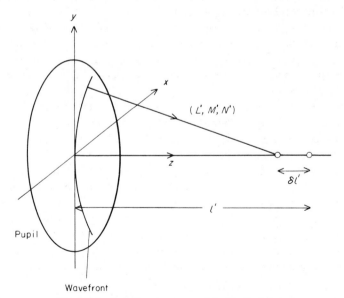

FIG. 9.8. Isoplanatism in a symmetrical optical system

ray with direction cosines (L', M', N'). For an off-axis object point in the y–z plane we take the rotation axis along the x-axis and so we have

$$\left.\begin{array}{l} \boldsymbol{p}' = (1, 0, 0) \\ \boldsymbol{D}' = (0, 0, l' - \delta l') \\ \boldsymbol{r}' = (L', M', N') \end{array}\right\}. \qquad (9.59)$$

The scalar triple product $\{\boldsymbol{p}', \boldsymbol{D}', \boldsymbol{r}'\} = S' \sin \theta'$ is easily found to be $-M'(l' - \delta l')$ and correspondingly in object space. We take $\varepsilon = \eta/l$, $\varepsilon' = \eta'/l'$ where η and η' are the (small) paraxial object and image heights. Then the change in aberration as the object point moves from the axis to a distance η from the axis is

$$\text{Linear coma} = -\frac{n'\eta'}{l'} M'(l' - \delta l') + \frac{n\eta}{l} Ml + C$$

$$= -H\left\{\frac{M'}{u'}\left(1 - \frac{\delta l'}{l'}\right) - \frac{M}{u}\right\}. \qquad (9.60)$$

The first expression strictly gives the change in aberration along the finite ray and we have to subtract from this the change in aberration along some standard ray; clearly this should be the axial ray, for which $M' = M = 0$ so that on carrying out the subtraction the constant C disappears and the second expression is obtained. It is easily seen that this is the same expression as in eqn (9.39). From this follow immediately the Staeble–Lihotzky condition and the Abbe sine condition as in Section 9.5.

A more sophisticated result is obtained if we consider the image formation by a pencil which is a finite distance from the axis of a symmetrical optical system. This is particularly relevant to the question mentioned in Section 9.4 of computing image formation of extended objects by convolution with the point spread function.

Let P'O' be the axis of a symmetrical optical system in image space, P' being the centre of the exit pupil and O' the Gaussian image point. Let η' be the (finite) coordinate of O'_1, the intersection of the principal ray with the Gaussian image plane, and let another ray, with direction cosines (L', M', N') meet the Gaussian image plane with transverse ray aberration components $(\delta\xi', \delta\eta')$; these points and rays are shown in Fig. 9.9. Now let the object height increase by $d\eta$; the principal ray from $\eta + d\eta$ still passes through the centre of the entrance pupil but since in general the pupil imagery is heavily aberrated this new principal ray will in general intersect the original principal ray at some point other than P', say P'_1.† The infinitesimal movement of the pencil in image space which we have to consider is therefore that which moves

---

† This point P'_1 is in fact the *tangential image* of the centre of the entrance pupil along the principal ray; see Section 9.8.

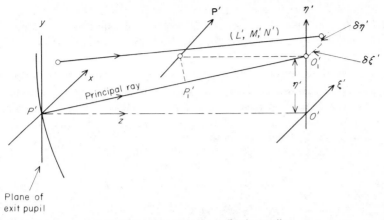

FIG. 9.9. Isoplanatism of an off-axis pencil

O_1' along the η-axis and P_1' along the principal ray. This axis p' must therefore be parallel to the x-axis and it must meet the y–z plane at the intersection of the perpendicular to the principal ray at P_1' and the perpendicular to the η-axis at O_1', as shown. The p-axis in object space is defined in a similar way but here P_1 is at the centre of the entrance pupil. We put $P_1'O_1' = t'$ and we take the vector $\boldsymbol{D}'$ from the point where the axis p' meets the y–z plane to the point where the general ray meets the Gaussian image plane. Then we have

$$
\left.
\begin{aligned}
\boldsymbol{p}' &= (1, 0, 0) \\
\boldsymbol{D}' &= (\delta\xi', \delta\eta', t'/\overline{N}') \\
\boldsymbol{r}' &= (L', M', N')
\end{aligned}
\right\}, \tag{9.61}
$$

where $\overline{N}'$ is the third direction cosine of the principal ray. Thus

$$
\{\boldsymbol{p}', \boldsymbol{D}', \boldsymbol{r}'\} = N' \, \delta\eta' - M't'/\overline{N}' \tag{9.62}
$$

and also $\varepsilon' = \overline{N}' \, \mathrm{d}\eta'/t'$, so that the change in aberration along the ray (L', M', N') is

$$
-\mathrm{d}\eta \left\{ n' \frac{\mathrm{d}\eta'}{\mathrm{d}\eta} (M' - \overline{N}'N' \, \delta\eta'/t') - nM \right\} + C.
$$

In order to eliminate the constant we subtract the change in path along the principal ray, giving finally for the aberration change corresponding to an object shift $\mathrm{d}\eta$,

$$
\mathrm{d}W = -\mathrm{d}\eta \left\{ n' \frac{\mathrm{d}\eta'}{\mathrm{d}\eta} (M' - \overline{M}' - \overline{N}'N' \, \delta\eta'/t') - n(M - \overline{M}) \right\}. \tag{9.63}
$$

In this expression the aberration change $\mathrm{d}W$ does not necessarily contain only coma-like terms; the unshifted off-axis pencil may have had aberrations of many types, including coma, and dW is a measure of the change in all these. The expression $d\eta'/d\eta$ is merely the local magnification of the system in the y–z section.

A rather different example is obtained by letting both axes p and p' move to an infinite distance in the general theorem of Section 9.6; if the rotations ε and ε' are made to decrease at the right rate we obtain in the limit infinitesimal *translations* δs and $\delta s'$ of the pencils. The theorem then gives as the necessary and sufficient condition that the image side pencil is translated through $\delta s'$ without distortion if the object side pencil is translated through δs,

$$n' \, \delta s' \cdot r' = n \, \delta s \cdot r + C + O(\sigma^2). \tag{9.64}$$

This result was given by T. Smith in 1922 (*Transactions of the Optical Society of London* **24**, 31 (1922–23)); he called it the *optical cosine law* and he deduced the Abbe sine condition from it. The optical cosine law appears at first sight to be a very general law but it is in fact not even possible to deduce Conrady's formula for offence against the siné condition (eqn (9.40)) from it; this is because the infinitesimal translation is not the most general movement possible for a pencil of rays: in effect it restricts us to the case where both the entrance and the exit pupil are at infinity. The only situation where the optical cosine law can be applied to derive results for pupils at a finite distance is when the pencils have no aberration, since a rotation of a spherical wavefront can always be replaced by a translation.

We next consider movement of the off-axis object point parallel to the axis, i.e. a longitudinal displacement. The diagram (Fig. 9.10) is similar to Fig. 9.9, but now we consider a displacement $d\zeta'$ of O'_1; thus the axis p' is obtained as the intersection of the perpendicular to the tangential image P'_1 and the η axis. We now have

$$\left.\begin{aligned}
p' &= (1, 0, 0) \\
D' &= (\delta\xi', \delta\eta' + t'/\overline{M}', 0) \\
r' &= (L', M', N') \\
\varepsilon' &= \overline{M}' \, d\zeta'/t'
\end{aligned}\right\}, \tag{9.65}$$

and we find for the change in aberration

$$dW = - \, d\zeta'\left\{n'\frac{d\zeta'}{d\zeta}\left(\overline{N}' - N' - \frac{\overline{M}'N'}{t'} \, \delta\eta'\right) - n(\overline{N} - N)\right\}. \tag{9.66}$$

If the pencil in this result is an axial pencil eqn (9.66) reduces to

$$dW = - \, d\zeta\left\{n'\frac{d\zeta'}{d\zeta}(1 - N') - n(1 - N)\right\}, \tag{9.67}$$

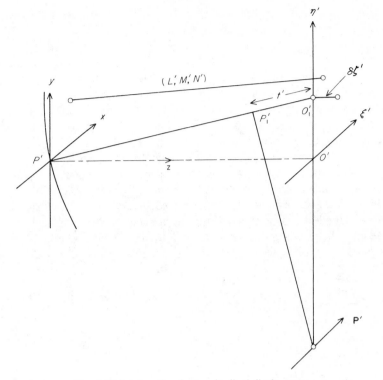

FIG. 9.10. Isoplanatism for longitudinal displacements

which is Herschel's condition for constancy of aberration with change of axial conjugates (J. F. W. Herschel (1821), *Phil. Trans. R. Soc.* **111**, 222).

In all the examples in this section ratios such as $d\zeta'/d\zeta$ occur; these are local magnification factors, i.e. particular forms of the ratio ε'/ε which occurs in the general theorem of Section 9.6. Their values can be obtained in terms of the generalized paraxial optics round a finite principal ray, the topic of Section 9.8.

Finally we note that if the rotation axes p and p' coincide with the axis of a symmetrical optical system the general isoplanatism theorem of Section 9.6 reduces to the skew invariant theorem (Chapter 6).

9.8   Optics round a finite principal ray

The principal ray of an off-axis or oblique pencil in a symmetrical optical system is a kind of axis or centre-line of the pencil. This rough concept can be usefully refined by developing a generalized paraxial optics of the region near the principal ray. We neglect powers greater than the square of the distance from the principal ray in expressing the shape of a wavefront of this pencil, just as in Gaussian optics; since the configuration of a finite principal ray in a symmetrical optical system has only a single plane of symmetry, the meridian plane containing the ray, the pencil can be astigmatic; however, since higher powers than squares are neglected aberrations such as coma and spherical aberration, which depend on the cube or fourth power of the aperture, are not taken account of. Thus we shall arrive at a method of calculating astigmatism along a finite principal ray.

The meridian plane containing the principal ray is also sometimes called the *tangential* plane; rays near the principal ray and lying in this plane are tangential rays and they produce a tangential focus. The plane through the principal ray and orthogonal to the meridian plane in each space is the *sagittal* plane but this in general changes at each refraction, as in Fig. 9.11; sometimes the term *sagittal section* is used to denote the ensemble of sagittal planes through the system which belong to a given principal ray. Rays near the principal ray which start off in the sagittal section remain in this section

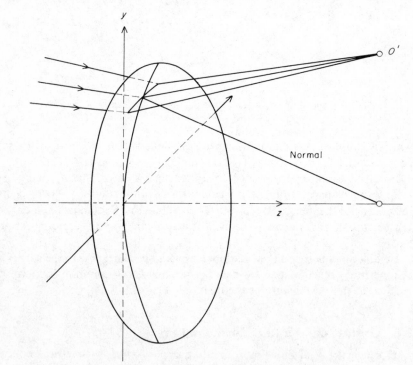

FIG. 9.11. The sagittal section; in each space it is a plane containing the principal ray and perpendicular to the meridian section

right through the system, to the approximation used in this theory, and they form the sagittal focus. These two foci are formed in every intermediate space of the system, possibly as virtual foci, and they can be tracked along the principal ray to the final image space. Thus we can obtain such diagrams as Fig. 9.12 which show the position of the two foci relative to the Gaussian image plane as a function of the image height; such a diagram is clearly an extension

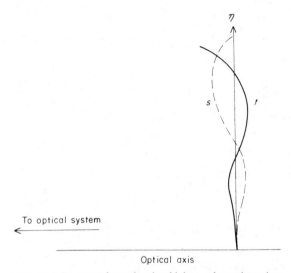

FIG. 9.12. Image surfaces showing higher order astigmatism

of, say, Fig. 7.29 to field angles beyond that for which the Seidel approximation is valid.

The formulae to be derived are usually called s- and t-tracing formulae; they are generally ascribed to Thomas Young as they were first given together by him in 1801 (*Phil. Trans. R. Soc.* **102**, 123–28) but special forms of the individual formulae had been discovered and published many years before that date.

In Fig. 9.13 r is a principal ray of an oblique pencil refracted through a spherical surface of curvature c with refractive indices n and n' on either side. Let T and T' be the tangential foci before and after refraction and let these be at distances t and t' from the point of incidence P. The angles of incidence are I and I' and another ray of the pencil r_1 in the tangential section meets the refracting surface at Q. Let PQ be denoted by h_t; then if h_t is small enough to make h_t^3 negligible the ray r_1 must meet r at T and T'. Snell's law gives

$$n \sin I = n' \sin I'$$

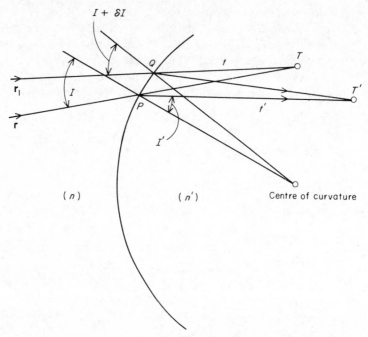

FIG. 9.13. Tangential foci

and differentiating this we have

$$n' \cos I' . \delta I' = n \cos I . \delta I. \tag{9.68}$$

If we wish this differentiation to apply to the shift from ray r to ray r_1 we have to put

$$\delta I = h_t \left(c - \frac{\cos I}{t} \right), \qquad \delta I' = h_t \left(c - \frac{\cos I'}{t'} \right)$$

and substituting in eqn (9.68) we have

$$n' \cos I' \left(c - \frac{\cos I'}{t'} \right) = n \cos I \left(c - \frac{\cos I}{t} \right)$$

or,

$$\frac{n' \cos^2 I'}{t'} - \frac{n \cos^2 I}{t} = c(n' \cos I' - n \cos I). \tag{9.69}$$

This is a conjugate distance equation for the tangential foci; it has a close resemblance to the paraxial conjugate distance equation (eqn (3.38)), to which it reduces if $\cos I$ and $\cos I'$ are put equal to unity.

We next obtain the corresponding relation for sagittal foci. In Fig. 9.14 let *r* again be the principal ray and let S be the sagittal focus in the incident space. We join S to the centre of curvature and call this line the auxiliary axis (it has, of course, nothing to do with the axis of symmetry of the complete system). Let the auxiliary axis meet the refracted part of the ray *r* at S'; if the three segments of ray shown are rotated about the auxiliary axis through any

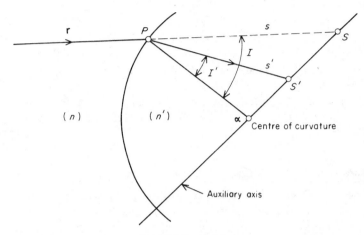

FIG. 9.14. Sagittal foci

angle they will still represent a possible configuration of incident and refracted ray at the surface, so that if this angle of rotation is small they will represent a ray close to the principal ray in the sagittal section. Thus we see that S' is the sagittal focus corresponding to S. To find an actual conjugate distance relation between PS = *s* and PS' = *s'* we note that

$$\left. \begin{array}{c} \dfrac{1}{c \sin (\alpha - I)} = \dfrac{s}{\sin \alpha} \\[2mm] \dfrac{1}{c \sin (\alpha - I')} = \dfrac{s'}{\sin \alpha} \end{array} \right\}. \tag{9.70}$$

If α is eliminated between these two equations we obtain

$$\frac{s \sin I}{1 - cs \cos I} = \frac{s' \sin I'}{1 - cs' \cos I'}, \tag{9.71}$$

or, from Snell's law,

$$\frac{s}{n(1 - cs \cos I)} = \frac{s'}{n'(1 - cs' \cos I')},$$

which may be rearranged to give

$$\frac{n'}{s'} - \frac{n}{s} = c(n' \cos I' - n \cos I),$$ (9.72)

the required sagittal conjugate distance equation.

Equations (9.69) and (9.72) are Thomas Young's astigmatism equations and they are used to trace the positions of the astigmatic foci along a finite principal ray. The raytrace data required are all available from the standard raytracing equations of Chapter 4. In particular we note that the generalized power as defined in Chapter 4 appears on the right-hand side of both astigmatism equations. Transfer between surfaces involves merely the distance D along the principal ray from surface to surface and this is also known from the raytrace data.

The modifications required for non-spherical surfaces of revolution are trivial: we simply have to use the appropriate principal curvature instead of the spherical curvature.

9.9 Astigmatism of quadrics of revolution

In this section we give as an example of the application of the formulae of Section 9.8 a proof of the result stated at the end of Chapter 8, that if the principal ray passes through a geometrical focus on a reflecting quadric of revolution there is no astigmatism for any conjugates on that ray. Let the

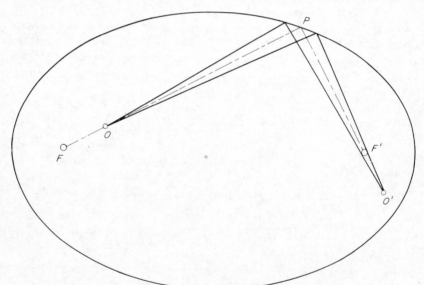

FIG. 9.15. Stigmatic reflection at a quadric

principal ray be as in Fig. 9.15 with conjugates $PO = l$ and $PO' = l'$ and let c_s and c_t be the curvatures of the quadric at P in the sagittal and tangential sections.

Then from eqns (9.69) and (9.72) we must have

$$\left.\begin{array}{l} \dfrac{1}{l} + \dfrac{1}{l'} = 2c_s \cos I \\[2mm] \dfrac{1}{l} + \dfrac{1}{l'} = \dfrac{2c_t}{\cos I} \end{array}\right\}, \qquad (9.73)$$

if there is no astigmatism for O and O' as conjugates. That is, we have

$$c_t = c_s \cos^2 I; \qquad (9.74)$$

but this condition is independent of the conjugates and so if it is true at one pair of conjugates O and O', it is true for all conjugates. But it is true for the conjugates F and F', since these are aberration-free, and therefore there is no astigmatism at all conjugates.

At any non-singular point on a general reflecting surface, if the principal curvatures c_s and c_t are of the same sign, an azimuth and an angle of incidence can always be found to satisfy eqn (9.74) and thus there is always one principal ray at any point for which there is no astigmatism at any conjugate.

10 Chromatic Aberration

10.1 Introduction: Historical aspects

The refractive index of any medium other than vacuum varies with wavelength. Thus the Gaussian and aberrational properties of any refracting optical system are functions of wavelength, i.e. chromatic *aberrations* exist. It is traditional to consider chromatic variations of Gaussian properties as aberrations as well as chromatic variations of true aberrations. This usage, although perhaps inaesthetic, has ample practical justification; in fact, many modern refracting systems intended for use over an appreciable range of wavelengths are ultimately limited in performance by chromatic effects, both Gaussian and higher order, rather than by the monochromatic aberrations which we have so far been considering. The history of astronomical telescope design provides a useful example. Sir Isaac Newton invented the reflecting telescope because he considered that it was impossible to correct the chromatic effects in singlet lenses by combining two to form what we now call an achromatic doublet; in effect he thought that the chromatic aberrations of all lenses were proportional to their powers, with the same constant of proportionality even for different glasses. Then, in the middle of the eighteenth century, John Dollond and Chester Moore Hall showed that Newton had been wrong and they made achromatic doublets. From this point larger and larger astronomical telescope doublet objectives were made, the largest ever being the famous Yerkes objective of 1 m diameter. However, as the objectives became larger and the design techniques became subtler it was found that an "achromatic" doublet was not quite free from chromatic aberration and a residual error, known as "secondary spectrum", appeared. This led to the re-introduction of reflecting objectives and these are now universally used for apertures exceeding 1 m. There are in fact other reasons why reflecting objectives are used exclusively for large telescopes but the secondary spectrum would be a sufficient reason if the others did not exist.[†]

A natural reaction to the above brief history is to say, "Well, if a doublet is limited by secondary spectrum why not try a more complex lens system, say a triplet?" The answer brings out the essential difference between the chromatic aberrations and the monochromatic aberrations; it is that adding more components generally has the effect of increasing chromatic residuals, unless we are content to accept a system with either unreasonably small relative aperture or unreasonably large monochromatic aberrations. This is because the variation of refractive index with wavelength, or *dispersion*, is a property of the material which is not governed by a simple physical law; in fact it is found necessary to use elaborate empirical formulae to represent the dispersion of optical glass with sufficient accuracy for optical design purposes and the constants and functional forms of the formulae are still matters for discussion. With monochromatic aberrations it seems possible, on the other hand, to improve the correction indefinitely by adding enough refracting or reflecting surfaces, although in practice we can reach a point where the complexity of the optical design is such that it could not be made accurately enough.

10.2 Longitudinal chromatic aberration and the achromatic doublet

The ideas suggested in the introduction are best illustrated by the example of the achromatic doublet. Suppose we place two thin lenses of powers K_1 and K_2 in contact; then from the results of Chapter 3 the power of the combination is

$$K = K_1 + K_2. \tag{10.1}$$

This relationship is by implication true for a certain wavelength λ at which the refractive indices of the materials of the two lenses are, say, n_1 and n_2. If now the wavelength is changed to $\lambda + \delta\lambda$ the refractive indices become $n_1 + \delta n_1$ and $n_2 + \delta n_2$ and we find, recalling from Chapter 3 that the power of a thin lens is $(n - 1)(c_1 - c_2)$,

$$\delta K = \frac{\delta n_1}{n_1 - 1} K_1 + \frac{\delta n_2}{n_2 - 1} K_2. \tag{10.2}$$

[†] For more details of the history of the achromat see L. C. Martin (1962), *J. R. microsc. Soc.* **80**, 247–58.

Thus the condition for zero change in power for the wavelength change $\delta\lambda$ is

$$\frac{\delta n_1}{n_1 - 1} K_1 + \frac{\delta n_2}{n_2 - 1} K_2 = 0. \tag{10.3}$$

If we postulate two different materials of known properties $(n, \delta n)$ it is clearly possible to solve eqns (10.1) and (10.3) as a pair of simultaneous equations for the powers K_1 and K_2 of the components of an achromatic doublet of power K.

The quantity $(n - 1)/\delta n$ of which the reciprocal appears in eqns (10.2) and (10.3) is usually denoted by V, or ν in the German literature; it has been called the *reciprocal dispersive power* and the *constringence* but is nowadays most usually called the V-value. It can be seen that if the V-values V_1 and V_2 of the glasses of the two components were equal then eqns (10.1) and (10.3) could only be solved for $K = 0$. This was Newton's mistake, to suppose that the V-values of all glasses were equal, although this quantity had not been explicitly defined in his time. In fact the values for optical glasses in the visible spectrum vary over a 4:1 range and achromatic doublets in this limited sense are easily made.

All optical media have increasing refractive indices with decreasing wavelength, except in regions of anomalous dispersion, and in such regions the

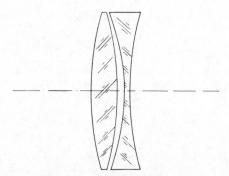

FIG. 10.1. An achromatic doublet

absorption losses are high enough to make the materials useless for optical design purposes. By convention the increment $\delta\lambda$ is taken as negative so all V-values are positive and thus any achromatic doublet designed according to eqns (10.1) and (10.3) has one positive and one negative component, as in Fig. 10.1.

What we have done so far is to make the power of the doublet the same at two different wavelengths λ and $\lambda + \delta\lambda$; the change in power $\delta'K$ at a third wavelength $\lambda + \delta'\lambda$ is, by analogy with eqn (10.2),

$$
\left.
\begin{aligned}
\delta'K &= \frac{\delta'n_1}{n_1 - 1} K_1 + \frac{\delta'n_2}{n_2 - 1} K_2 \\
&= \frac{\delta'n_1}{\delta n_1} \frac{K_1}{V_1} + \frac{\delta'n_2}{\delta n_2} \frac{K_2}{V_2}
\end{aligned}
\right\}.
\tag{10.4}
$$

Clearly if $\delta'K$ is to vanish we must have

$$
\frac{\delta'n_1}{\delta n_1} = \frac{\delta'n_2}{\delta n_2}.
\tag{10.5}
$$

These quantities are called relative partial dispersions and they are in fact *not* equal for different glasses; thus there is a residual variation of power with wavelength even with a so-called achromatic doublet, but the variation is much less than with a singlet lens of either material and the same power. To see the nature of this residual variation, the *secondary spectrum* mentioned in Section 10.1, and its order of magnitude, it will be useful to examine the dispersion of optical materials in more detail and we do this in the next section.

10.3 Dispersion of optical materials

When we speak of optical materials in connection with aberration theory we generally mean the optical glasses; these are the materials almost invariably used in the visible spectrum and it is in this region that the most sophisticated optical designs have been carried out; we shall therefore consider mainly optical glasses. It follows from Section 10.2 that the V-values and the relative partial dispersions are important parameters for chromatic correction and clearly the refractive index itself at some mean wavelength is important for Gaussian design and for monochromatic aberrations. The manufacturers of optical glass list these data for their glasses and a catalogue of optical glasses is therefore a prime source of numerical data for optical design. Refractive indices are generally given to the fifth decimal and relative to air at normal temperature and pressure and dispersions to an accuracy equivalent to about the sixth decimal in index. To achieve this precision it is necessary to use sharp spectral lines or laser lines for measurement and the index values are given for about a dozen lines ranging from $0.365\ \mu m$ in the ultraviolet to $1.014\ \mu m$ in the infrared (both mercury lines). The mean wavelength is conventionally taken as the helium d line ($0.5876\ \mu m$) and the wavelength range for specifying the V-value is, again by convention, usually from the hydrogen

C line (0·6563 μm) to the hydrogen F line (0·4861 μm). Thus the V-value as normally given is defined by

$$V = \frac{n_d - 1}{n_F - n_C}.$$

(10.6)

It should be stressed that the wavelengths used are conventionally chosen; they correspond roughly to the middle of the visible spectrum (d) and the red and blue extremes, C and F, beyond which the visual effect of the light is usually insignificant. However, the actual wavelengths used to define the V-values for a particular problem must depend on the nature of the light source and detector to be used.

In order to display the main properties of glasses in graphical form the glass catalogues use a graph in which V is plotted as abscissa and n as ordinate, the $n–V$ diagram; each optical glass appears as a point on the diagram and groups of glasses in different regions are given generic names (hard crown, dense flint, lanthanum crown, etc.), which are mostly of historical origin. Figure 10.2 shows the form of the $n–V$ diagram with an indication of the regions where points for actual glasses can be found. The available refractive indices range from 1·45 to 1·95 and the V-values (C to F) from about 80 to 20.

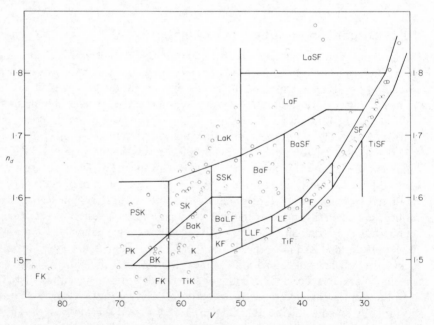

FIG. 10.2. The glass chart; some points representing frequently melted glass types and also some of the newer glasses with extreme properties, taken from the Schott catalogue

Each glass type is denoted by the maker's abbreviation for the group, followed by an arbitrary serial number. Thus the borosilicate crown made by Schott with $n_d = 1.51680$ and $V = 64.29$ is BK7. Alternatively a more informative method is to use the six digit number of which the first three digits are the first three decimal digits of the refractive index and the last three digits are the three most significant digits of the V-values; according to this system BK7 would be 517643.

The mathematical form of the dependence of refractive index on the wavelength (or frequency) of light is quite complicated when a precision of 0.00001 in index is in question. The Sellmeier type of dispersion formula,

$$n^2 - 1 = \sum_i \frac{c_i \lambda^2}{\lambda^2 - \lambda_i^2} \qquad (10.7)$$

has perhaps the best claim to represent a convincing physical theory, the λ_i being the wavelengths of absorption band maxima, but in fact it does not represent the experimental data very well for optical glasses. Other formulae which have been used are due to Cauchy,

$$n = A + \sum_i \frac{B_i}{\lambda_i^2}, \qquad (10.8)$$

Hartmann,

$$n = n_0 + \frac{c}{(\lambda - \lambda_0)^{1.2}} \qquad (10.9)$$

and Conrady,

$$n = n_0 + \frac{a}{\lambda} + \frac{b}{\lambda^{3/2}}. \qquad (10.10)$$

However, none of these formulae fits the glasses now available over the full wavelength range from near ultraviolet to near infrared to the accuracy of the measurements. The best fit is obtained by a formula which was given in the 1967 Schott optical glass catalogue and which is now used by other manufacturers as well. The formula is

$$n^2 = A_0 + A_1 \lambda^2 + \frac{A_2}{\lambda^2} + \frac{A_3}{\lambda^4} + \frac{A_4}{\lambda^6} + \frac{A_5}{\lambda^8}; \qquad (10.11)$$

the constants A_0 to A_5 are listed for each glass type and the formula seems to be reliable for interpolation to better than 0.00001 in index. The values of the six constants vary considerably between glasses and thus the general shapes of all dispersion curves will be different, provided they are examined in enough detail to see index variations of 0.00001. Thus we must expect that if an achromat is made according to eqns (10.1) and (10.3) there will inevitably be a residual mismatch of the dispersions, leading to secondary spectrum. From eqns (10.4) and (10.5) it can be seen that in order to eliminate secondary spectrum we should need to find a pair of glasses with different V-values but the *same* relative partial dispersions.

We define a general relative partial dispersion P_{λ_1, λ_2} as

$$P_{\lambda_1, \lambda_2} = \frac{n_{\lambda 1} - n_{\lambda 2}}{n_F - N_C};$$

(10.12)

thus if this quantity were the same for two glasses, an achromat designed for the range C to F would also have the same power for λ_1 as it had for λ_2. We can seek for suitable pairs of glasses by plotting P against V, as in Fig. 10.3, which shows a selection of points from the Schott glass catalogue; the values of λ_1 and λ_2 are for the hydrogen F line and the d line of helium (0·5876 μm). We see that there is roughly a linear relation between P and V, a fact first noted by Abbe; pairs of glasses of which the representative points lie on a horizontal line would be needed to make an achromat with reduced secondary spectrum but in order to keep the powers of the individual components to reasonable values so as to reduce monochromatic aberrations the V-values should be as different as possible.

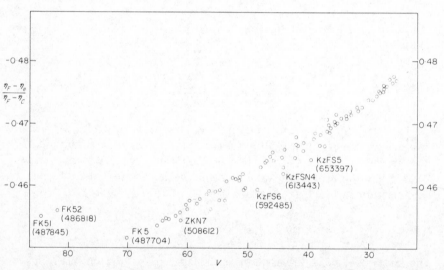

FIG. 10.3. The Abbe plot of relative partial dispersion against V-value. The points are for some glasses in the Schott catalogue; some of the glasses lying apart from the general run have type symbols (KzFS, ZKN) which do not appear in Fig. 10.2; these types are given a special designation to indicate their different relative partial dispersions but their representative points do not fall in a separate part of the glass chart

Figure 10.4 shows the residual lack of achromatism (secondary spectrum) in thin lens achromatic doublets. A typical pair of glasses would be K5 (522595) and KzFSN4, (613443); for this pair it can be seen that the residual variation of power from C to F is about 0.03% and this is typical of what can be expected of almost any pair of glasses chosen at random. If instead of K5 we choose FK51, (487845) the secondary spectrum is seen to be halved, as

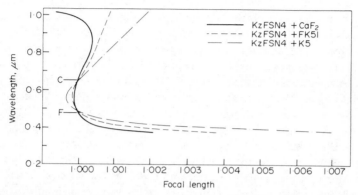

FIG. 10.4. The secondary spectrum of achromats; the KzFSN4 is the "flint", i.e. the material of higher dispersion. The graphs are calculated for a doublet of unit power at C and F

would be expected from Fig. 10.3, where FK51 can be seen to lie well to the left of the mean straight line. The designation FK stands for fluor-crown, meaning a glass having characteristics approaching those of fluorite (CaF_2). An achromat made with pure fluorite as the positive component would show even smaller secondary spectrum, as in the third graph of Fig. 10.4; however this material is expensive, unobtainable in large pieces and difficult to work to a good polish, so that it is not often used; in fact the main use is in microscope objectives, which can be made to have the very high degree of achromatism known as apochromatism, and for ultraviolet transmitting achromats, where it is generally used in combination with fused silica for the negative element. The fluor-crowns are not often used in simple achromatic doublets for various practical reasons concerned with cost, chemical stability and homogeneity.

We can now see some justification for the statement in Section 10.1 that the secondary spectrum cannot in practice be much reduced by adding more components or by other means. To take the simplest possible case, the achromatic doublet might be required as a collimator for a point source, e.g. an illuminated pinhole, and then the only monochromatic aberration to be corrected would be spherical aberration; if glasses with more nearly equal relative partial dispersions are chosen from Fig. 10.3 their V-values will be very close and consequently the individual powers of the elements would be numerically large compared to the total power of the doublet. This sets an upper limit to the relative aperture of the system if spherical aberration is to be kept down.

We have discussed at some length in this section the effects of dispersion on chromatic aberration in a purely numerical way. This discussion has shown, as stated in Section 10.1, that it is not profitable to attempt to deal with chromatic aberration by elaborate mathematical analysis since the dispersions of glasses do not conform to reasonably simple mathematical forms. This, and the resultant fact that no glasses with reasonably different V-values have the same relative partial dispersions, are occasionally graphically summed up as the "irrationality of dispersion": this phrase originally meant simply that the relative partial dispersions, i.e. the *ratios*, differed between glasses but nowadays it carries appropriate overtones suggesting the intransigence of optical glass.

Sometimes systems are needed where the wavelength range to be used is relatively small. For example, the light source may be a cathode ray tube with a phosphor emitting a narrow wavelength range, or the detector may respond only over a narrow wavelength range. In such cases a quadratic approximation to the dispersion may be adequate and it may be possible to treat chromatic aberration more analytically than is customary for wide-band systems. In extreme cases, e.g. systems designed for use with lasers, no chromatic correction is needed.

10.4 Chromatic aberration for finite rays; the Conrady formula

If we examine chromatic effects in detail we find, naturally, that all aberrations vary with wavelength in a refracting system. Exact formulae for the chromatic variations of even the Seidel aberration sums are very cumbersome, excepting, of course, for S_{IV}, the Petzval sum. It might be thought that one could obtain formulae for chromatic variations of the aberrations simply by differentiating the Seidel sum formulae with respect to refractive index, but in fact the other quantities which appear, A, $\bar{A}$, h and u, also have non-zero first derivatives with respect to refractive index; thus the results would in general be too unwieldy for general use.

It is, however, a trivial matter simply to trace more rays, both paraxial and finite, in a selection of wavelengths, in order to obtain full details of the behaviour of the system through the spectrum. The results, expressed either as transverse ray chromatic aberration or as wavefront aberration in the new wavelengths, provide full details of chromatic residuals. This method is universally adopted when reasonably fast computers are available.

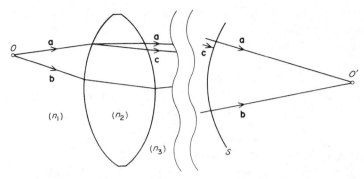

FIG. 10.5. The Conrady formula for chromatic aberration of finite rays

There is in addition an interesting formula† which can be applied to finite rays to give an approximation to the chromatic aberration which is linear with wavelength. In Fig. 10.5 *a* and *b* are two finite rays from a point O in object space which intersect at O′ in image space; the optical system is not necessarily symmetrical. If we construct a reference sphere S centred on O′ the optical path difference between the two rays is

$$W = \sum n(D_b - D_a) \tag{10.13}$$

where D_a is a typical segment of ray *a* and the summation is taken over all segments from O to the reference sphere S. Let the rays *a* and *b* as drawn be for wavelength λ; if the wavelength is changed to $\lambda + \delta\lambda$ then the ray *a* will change its path after the first refraction to *c* as indicated and similarly for *b*. The new optical path difference can be written

$$W + \delta_\lambda W = \sum (n + \delta n)(D_b + \delta D_b - D_a - \delta D_a)$$

so that

$$\delta_\lambda W = \sum n(\delta D_b - \delta D_a) + \sum \delta n(D_b - D_a), \tag{10.14}$$

if we neglect second order terms like $\delta n\, \delta D_b$. Now $\sum n\, \delta D_b$ is the change in optical path length from O to S in wavelength λ when instead of the physically possible ray *a* we go to the neighbouring path *c*, which is not a physical ray path;‡ but by Fermat's principle (Chapter 2) the optical path length along ray *a* is stationary so that $\sum n\, \delta D_b$ is a second order small quantity and likewise $\sum n\, \delta D_a$. Thus we have the result that the first order chromatic change in optical path length is

$$\delta_\lambda W = \sum \delta n(D_b - D_a). \tag{10.15}$$

† A. E. Conrady (1904), *Mon. Not. R. astr. Soc.* **64**, 182–188; 458–560; R. A. Herman, "A Treatise on Geometrical Optics", Cambridge University Press, 1900 (Article 169).
‡ The key point of this argument is that path *c* is a physically possible path in wavelength $\lambda + \delta\lambda$ but not in wavelength λ.

In particular if O and O' are finite axial conjugates and one of the rays is the axis we have for the chromatic aberration of the finite axial ray

$$\sum \delta n(D - d)\qquad(10.16)$$

where d is a typical axial component thickness. From the form of eqn (10.16) the result of this section is usually called rather uneuphoniously "the Conrady D minus d formula". It applies to the path difference between any pair of intersecting finite rays if the reference sphere is taken to be centred at the point of intersection in image space.† It is also possible to paraxialize the formula and so obtain, for example, a very simple derivation for the chromatic aberration of a thin lens. Finally we note that in computing the terms $D_b - D_a$ in eqn (10.15) it is not necessary to take account of air spaces since δn is only non-zero in dispersive media.

10.5   Expressions for the primary chromatic aberrations

As noted in Section 10.1 it is usual to talk of the chromatic effects on Gaussian properties as the primary or first order chromatic aberrations. We have already seen in Section 10.2 how one of these arises, longitudinal chromatic aberration, corresponding to a chromatic change in focusing distance or back

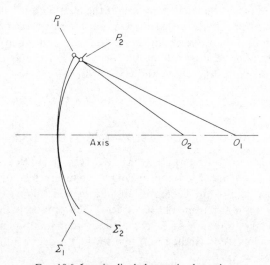

FIG. 10.6. Longitudinal chromatic aberration

† If the reference sphere is not centred at this point then the rays do not meet it normally and the shift in ray paths produces a first order change in optical path. For small displacements of the centre of the reference sphere this error would probably be numerically small even if mathematically significant.

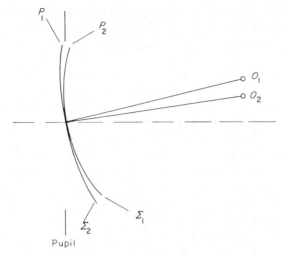

FIG. 10.7. Transverse chromatic aberration

focal length of an optical system; clearly there must be another effect, chromatic change of magnification or transverse chromatic aberration, corresponding roughly to a chromatic change in equivalent focal length. In this section we examine the latter in more detail and we derive surface-by-surface expressions for both aberrations.

Let Σ_1 in Fig. 10.6 be a paraxial wavefront from the axial object point in the pupil in an intermediate space of an optical system. We suppose Σ_1 to be formed by light of wavelength λ. Let Σ_2 be another wavefront from the same object point but formed by light of wavelength $\lambda + \delta\lambda$. If Σ_1 and Σ_2 converge respectively to the intermediate image points O_1 and O_2 then the distance O_1O_2 is a measure of the longitudinal chromatic aberration. However, just as for monochromatic aberrations, it turns out to be simpler to deal with optical path differences, since effects can be summed through the system. We assume Σ_1 and Σ_2 touch on the axis and we take points P_1 and P_2 where the two wavefronts meet a ray from the edge of the pupil to O_1. Then the optical path length $[P_1P_2]$ is taken as a measure of the longitudinal chromatic aberration.†
We are in the Gaussian optics region here, so that both wavefronts are portions of spheres and we are dealing with a chromatic shift of focus.

We can define transverse chromatic aberration as an optical path difference similarly. In Fig. 10.7 let Σ_1 be an off-axis paraxial wavefront converging to the

† The optical path is the geometrical distance multiplied by the refractive index, so the question arises, which index to use, n or $n + \delta n$? In fact, it is assumed that $\delta\lambda$ is mathematically small and then $P_1P_2.\delta n$ would be an order of magnitude smaller than $P_1P_2.n$. Thus we ignore the difference and use the original refractive index.

intermediate off-axis image O_1 formed in wavelength λ. It may happen that for wavelength $\lambda + \delta\lambda$ the image is formed at O_2 by the wavefront Σ_2 and we then use the optical path difference $[P_1P_2]$ at the edge of the pupil as a measure of transverse chromatic aberration. It is a chromatic change of magnification. In practice there would generally be some longitudinal chromatic aberration appearing at the same time but we have omitted this from the figure. We should expect the longitudinal chromatic aberration to be constant over the field in Gaussian approximation whereas the transverse chromatic aberration should depend linearly on field angle or height. These predictions are verified by the formulae obtained below.

In Chapter 3, Section 3.7, we obtained the conjugate distance equation for a single surface by the condition that the difference between the optical paths along the axis and along a ray should be zero. From eqns (3.36) and (3.37) this path difference is

$$W = \tfrac{1}{2}h^2 \,\Delta n\left(c - \frac{1}{l}\right) \tag{10.17}$$

and the conjugate distance equation is obtained by equating the right-hand side to zero. If we now differentiate eqn (10.17) with respect to wavelength we obtain

$$\delta_\lambda W = \tfrac{1}{2}h^2 \,\Delta\,\delta n\left(c - \frac{1}{l}\right)$$

$$= \tfrac{1}{2}\,Ah\,\Delta\,\frac{\delta n}{n}, \tag{10.18}$$

recalling the definition of the refraction invariant A (eqn (6.2)). Thus we have a simple expression for the contribution of a single surface to the longitudinal chromatic aberration as a wavefront aberration. It is usual to denote the sum of the expressions $Ah\Delta(\delta n/n)$ by C_1 by analogy with the Seidel sums and then the total longitudinal colour as wavefront aberration is

$$\tfrac{1}{2}C_1 = \tfrac{1}{2}\sum Ah\,\Delta\!\left(\frac{\delta n}{n}\right). \tag{10.19}$$

The derivation of the surface contribution to transverse colour is even simpler. Figure 10.8 shows a refracting surface with an off-axis pencil; the (paraxial) principal ray is incident at $\bar{P}$ at angles i and i' before and after refraction and the second ray of the pencil shown meets the surface at Q. Let P and P′ be points on this ray on the incident and refracted wavefronts which pass through $\bar{P}$. The path difference [PP′] is

$$W = n'.P'Q - n.PQ \tag{10.20}$$

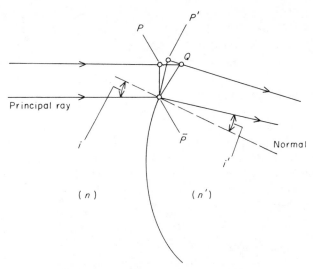

FIG. 10.8. Calculating the transverse colour

and this must vanish, by Fermat's principle, for physically possible rays. The paraxial approximation eqn (10.20) becomes

$$W = h(n'i' - ni) \qquad (10.21)$$

where h is the incidence height of the second ray. Differentiating with respect to wavelength we have immediately for the contribution to transverse colour

$$\bar{A}h \, \Delta\left(\frac{\delta n}{n}\right) \qquad (10.22)$$

The sum of such expressions over the system is denoted by C_{II}. The sums C_I and C_{II} are usually coupled with the Seidel sums as constituting a group of primary aberrations.

In the above derivations of the formulae for C_I and C_{II} it will be noted that when we differentiated with respect to wavelength we ignored variations of A, h, i etc. with wavelength and considered only the effects of explicit changes in n. In fact, of course, all the paraxial quantities must change with wavelength, but we can use the same argument as in Section 10.4 to show that these effects are small and they can therefore be neglected in first approximation. In fact it is possible to paraxialize the Conrady formula (eqn (10.15)) and arrive at the same expressions for C_I and C_{II}.

The increment δn is the change in refractive index of each medium corresponding to a chosen wavelength change $\delta \lambda$. The derivations above and the discussions in the earlier sections of this chapter would suggest that $\delta\lambda$ should be small, but in practice it is usually taken as the full wavelength range over which the system is to be used. This apparently inconsistent procedure can be partly justified by arguments similar to those used to justify paraxial calculations at full aperture and field (Section 3.10). Consider our previous example of an achromatic collimator (Section 10.3). If $\delta\lambda$ is taken as the wavelength range from F to C then the achromatizing equations will ensure that, within the thin lens approximation, the power of the doublet is *precisely* the same for the F and C wavelengths; elsewhere it will vary, giving the secondary spectrum already described. Similarly if the transverse colour is made zero between the F and C wavelengths there will be a residual secondary spectrum of transverse colour and this will cause all wavelengths between F and C to focus at a greater (or smaller) image height.

10.6 Stop-shift effects

In considering chromatic stop-shift effects we assume, as in Chapter 8, that when the aperture stop is moved along the optical axis, its diameter is changed so as to keep the size of the axial pencil constant. Thus again we can specify the stop-shift in terms of the increment $\varepsilon = H\,\delta E$ to the eccentricity parameter $HE\,(=\bar{h}/h)$. As for S_1, C_1 is independent of the stop position, since the formula (eqn 10.19)) contains only variables of the axial pencil. For C_{II} we use eqn (8.73) for the change in $\bar{A}$ and we obtain immediately

$$\delta C_{II} = \varepsilon C_1. \tag{10.23}$$

It follows that the transverse colour is independent of the stop position if there is zero longitudinal colour.

10.7 Ray aberration expressions for C_1 and C_{II}

We can obtain ray aberration expressions for the primary chromatic aberration sums C_1 and C_{II} by the same methods as were used in Section 8.4 for getting ray aberrations from the Seidel sums.

For longitudinal chromatic aberration the axial displacement of the focus, i.e. the chromatic focal shift, is

$$\delta_\lambda l'_k = \frac{1}{n'_k u'^2_k} C_1, \tag{10.24}$$

where n'_k and u'_k are the refractive index and convergence angle of the paraxial ray in image space. A positive value of $\delta_\lambda l'_k$ corresponds to a focal shift to the left with increasing refractive index, i.e. shorter wavelengths.

The chromatic change in image height of the principal ray, i.e. transverse displacement of the image, is

$$\delta_\lambda \eta'_k = -\frac{1}{n'_k u'_k} C_{II} \tag{10.25}$$

and a positive value indicates a decrease in image size with decreasing wavelength. The fractional transverse colour, analogous to fractional distortion as in eqn (8.63), is

$$\frac{\delta_\lambda \eta'_k}{\eta'_k} = -\frac{1}{H} C_{II}. \tag{10.26}$$

10.8 Some examples

The achromatic doublet, discussed in Section 10.2, is an example which brings out many important aspects of chromatic aberration. A very different example is the Huygens eyepiece. If two thin lenses of powers K_1 and K_2 and the same V-values are separated by a distance d the power of the combination is, from Chapter 3,

$$K = K_1 + K_2 - dK_1K_2. \tag{10.27}$$

The condition that K shall be unchanged by a wavelength shift $\delta\lambda$ is easily found to be

$$\frac{K_1 + K_2}{V} - \frac{2dK_1K_2}{V} = 0; \tag{10.28}$$

thus if

$$d = \frac{1}{2}\left(\frac{1}{K_1} + \frac{1}{K_2}\right) \tag{10.29}$$

the power is constant with wavelength; this is nominally the condition fulfilled by the Huygens eyepiece (Fig. 10.9), i.e. that the separation should be the mean of the focal lengths. Under this condition if the entrance pupil is at

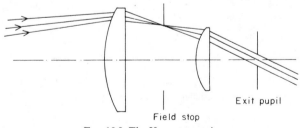

Field stop

Exit pupil

FIG. 10.9. The Huygens eyepiece

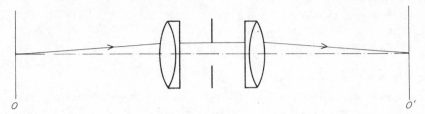

FIG. 10.10. A system with complete symmetry about the plane of the aperture stop, including magnification -1; it has no transverse colour (also it has no coma or distortion)

infinity and if the image is at infinity, as indicated in the figure, there will be no transverse chromatic aberration in the image. There *will* be longitudinal chromatic aberration since the condition does not ensure that the positions of the principal planes of the combination do not change with wavelength. It is worth noting that although this result is very easily obtained in the above way it is quite laborious to obtain it by calculating the transverse colour of the system by means of eqn (10.22) or by means of the thin-lens equations in Chapter 12; the reason is that it is necessary to put in explicitly the position of the pupil to get the calculation to come out.

If a system is completely symmetrical about its aperture stop, including the conjugates, as in Fig. 10.10, it has zero transverse colour; this is because transverse colour depends on an odd power of the field angle and the contributions of the two halves of the system therefore cancel.† If the conjugates are not completely symmetrical the residual transverse colour is usually small.

If it is necessary to control or vary the chromatic correction without changing the monochromatic aberrations this may sometimes be done by changing the glass of a component, keeping the same refractive index as the original but selecting a different V-value. However, it may not always be possible to get precisely the right change of V-value wanted for this purpose,

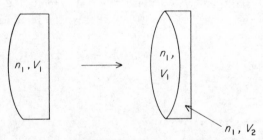

FIG. 10.11. Inserting a ghost contact surface to control colour without affecting the monochromatic aberrations

† The same is true, by similar reasoning, for coma and distortion, as noted in Section 8.8

since not many glasses lie accurately on the same horizontal lines in Fig. 10.2; in this case we can get more flexibility by putting in a "ghost" contact surface, as in Fig. 10.11, where the curvature can be varied continuously to control chromatic aberration.

Allied to this is the concept of a doublet of given V-value. It may happen that a thin-lens design calls for a lens in a certain position of power K and a V-value not obtainable with known glasses. In this case we can make up this thin lens as a doublet with individual components K_1 and K_2 and of glasses with V-values V_1 and V_2. The total chromatic change of power is, as in Section 10.2,

$$\delta K = \frac{K_1}{V_1} + \frac{K_2}{V_2}$$

and if δK is *not* zero we have a system of which the effective V-value is

$$V = \frac{K_1 + K_2}{\dfrac{K_1}{V_1} + \dfrac{K_2}{V_2}}. \tag{10.30}$$

11 Primary Aberrations of Unsymmetrical Systems and of Holographic Optical Elements

This chapter contains a sketch of the primary aberration theory of some lens and mirror systems which are not axisymmetric, diffraction gratings and holographic optical elements. The discussion of lenses and mirrors is restricted to systems in which all the cylinder axes intersect a single line and are either parallel or perpendicular to it. The discussion of holographic optical elements is limited to holograms formed on surfaces which are either plane or are in the form of quadrics (conicoids) of revolution, since these are at present the only cases for which reasonably manageable results are available.

11.1 Cylindrical systems

This section is based on work by C. G. Wynne (*Proc. Phys. Soc.* **LXVIIB**, 529, 1954); we give only an outline of Wynne's treatment, omitting details of the algebra and of the justifications for neglecting terms of higher order.

11.1.1 Gaussian optics of cylindrical systems

A cylindrical surface, refracting or reflecting, has power in one section and behaves like a plane surface in the perpendicular section. Thus in the systems we are to consider a single surface has axial astigmatism. However, we could have a system of crossed cylinders or cylinders combined with spherical elements which is stigmatic on axis for a certain object position. Such a system would in general have different magnifications in two perpendicular transverse directions, i.e. it would be anamorphic. A simple illustration is given in

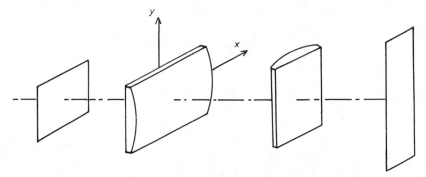

FIG. 11.1. An anamorphic optical system

Fig. 11.1, which shows crossed positive cylindrical lenses giving a stigmatic image (in Gaussian approximation) with magnifications of -2 and $-0\cdot5$ in the x and y sections.

In an anamorphic system the Gaussian optics of the two principal sections is treated separately, using the methods of Chapter 3. Since the longitudinal magnifications in the two sections differ we should not expect to find that the system is stigmatic for all conjugates and we can show that in general there will be just two pairs of conjugates for which this is true. For, if K_1 and K_2 are the powers of the system in the two sections the conditions for stigmatic imagery are

$$\left. \begin{aligned} \frac{1}{l_1'} - \frac{1}{l_1} &= K_1 \\[2mm] \frac{1}{l_1' - d'} - \frac{1}{l_1 - d} &= K_2 \end{aligned} \right\} \tag{11.1}$$

where d and d' are the distances between the respective principal planes in the two sections. Then if l_1 is eliminated between these two equations we obtain a quadratic in l_1', showing that there are just two solutions.

11.2   Primary aberrations of anamorphic systems

We assume initially that all the surfaces have their curvatures in the y section. Since a given surface contributes axial astigmatism there is not a unique Gaussian image plane in each intermediate space; this complicates the summation of aberration terms surface by surface because if we were to refer the contributions of each surface to one intermediate image plane there would be second degree as well as fourth degree terms. C. G. Wynne circumvented

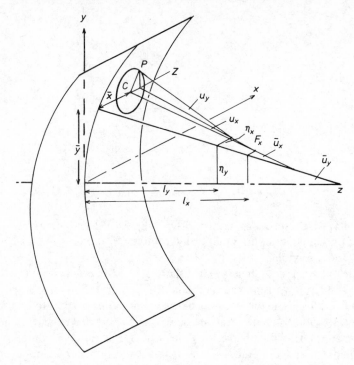

FIG. 11.2. Notation for primary aberrations at a cylindrical surface

this difficulty by summing fourth degree contributions for the two sections separately and allowing them to be combined only in a space, usually the final image space, where the Gaussian image is stigmatic.

Figure 11.2 shows a single cylindrical surface of radius of curvature r in the y section. The usual Gaussian optics holds in this section, with

$$\Delta n\left(\frac{1}{r} - \frac{1}{l_y}\right) = 0, \qquad nu_y\eta_y = H_y = \text{const.} \qquad (11.2)$$

Also in the x section, since there is zero power,

$$\Delta\left(\frac{n}{l_x}\right) = 0. \qquad (11.3)$$

Figure 11.2 shows a paraxial (skew) principal ray meeting the surface at C and the other rays of the pencil meet the surface within the elliptical area indicated. The distances l_x and l_y as computed from eqns (11.2) and (11.3) are indicated and we mark on the principal ray the point F_x which is the intersection of the principal ray with the plane $z = l_x$. This point F_x is a distance η_x from the y–z plane. Similarly the principal ray meets the plane $z = l_y$ at a distance η_y from the x–z plane. We take an outer ray of the pencil which meets the surface at P and another ray, through Z, having the same x coordinate as P and passing through the same cylinder generator as C. Then to the approximation needed the ray through Z meets the principal ray at F_x and the ray through P meets the ray through Z at F_y, the intersection with the plane $z = l_y$. Thus F_x and F_y are at intermediate astigmatic focal lines.

The optical path difference W_y introduced by the surface between the rays through P and Z referred to the focus F_y is a fourth degree quantity and consequently it may be summed surface-by-surface; the same applies to the optical path difference W_x between the rays through Z and C referrred to the focus F_x. Then in any space where F_x and F_y coincide, i.e. where there is axial stigmatism, W_x and W_y may be added to give the total primary aberration of the ray through P relative to the principal ray.

Wynne's results are expressed in terms of the data of two paraxial rays traced through a system of spherical surfaces having the same curvatures, etc., as the cylinders; thus we have the usual h, $\bar{h}$, etc. at all the surfaces. Also he notes that the pupil is circular in a space where the axial image is stigmatic and in such a space we can define $ZP/h = \sin \phi_0$ (ZP/h is invariant through the system although the azimuth of P changes) and $\eta_y/\eta = \sin \psi_0$, where η refers to the paraxial principal ray mentioned above. Finally we can trace two paraxial rays through a system of plane surfaces corresponding to the x–z section and obtain U_x, $\bar{U}_x$ etc., and it follows that $u_x/U_x = \cos \phi_0$ and $\bar{u}_x/\bar{U}_x = \cos \psi_0$. Using all these definitions Wynne's method gives for W_y

$$
\begin{aligned}
W_y = \{ & \tfrac{1}{8} \sin^4 \phi_0 \, S_{Iy} + \tfrac{1}{2} \sin^3 \phi_0 \sin \psi_0 \, S_{IIy} \\
& + \tfrac{1}{4} \sin^2 \phi_0 \sin^2 \psi_0 (3 S_{IIIy} + S_{IVy}) \\
& + \tfrac{1}{2} \sin \phi_0 \sin^3 \psi_0 S_{Vy} \} + n^2 \{ U_x \cos \phi_0 + \bar{U}_x \cos \psi_0 \}^2 \\
& \times \{ \sin^2 \phi_0 Ay + \sin \phi_0 [\sin \psi_0 \bar{A}y + (1 - \sin \psi_0) A\bar{y}] \} \Delta \left(\frac{1}{n^2} \right)
\end{aligned}
$$

(11.4)

In the above expression S_{Iy}, S_{IIy} etc. are the Seidel aberration terms corresponding to the paraxial rays traced in the y section through spherical surfaces.

For the second component of aberration, W_x, the result is

$$W_x = \{\tfrac{1}{8}\cos^4 \phi_0 S_{\mathrm{I}x} + \tfrac{1}{2}\cos^3 \phi_0 \cos \psi_0 S_{\mathrm{II}x}$$
$$+ \tfrac{3}{4}\cos^2 \phi_0 \cos^2 \psi_0 S_{\mathrm{III}x}$$
$$+ \tfrac{1}{2}\cos \phi_0 \cos^3 \psi_0 S_{\mathrm{V}x}\} + \tfrac{1}{4}nl_x U_x \cos \phi_0$$
$$\times \{U_x \cos \phi_0 + 2\bar{U}_x \cos \psi_0\}$$
$$\times \left\{\bar{A}^2 \Delta\left(\frac{1}{n^2}\right) - 2\frac{\bar{A}}{H^2}\bar{y}S_{\mathrm{IV}y} + \frac{n\bar{y}^2}{l_x}\cdot\frac{S_{\mathrm{II}y}}{H^2}\right\}\sin^2 \psi_0 \qquad (11.5)$$

where $S_{\mathrm{I}x}$, $S_{\mathrm{II}x}$ etc. are the Seidel aberration terms for the paraxial rays traced through the system of plane surfaces and the other symbols have the same meanings as before.

As explained above, W_x and W_y may be added in the final image space or wherever the Gaussian imagery is stigmatic to give the total of fourth degree aberration terms.

If there are crossed cylindrical surfaces there is an additional complication: in the space in which the transition occurs it is necessary to add a transverse focal shift to W_y in order to refer it to a point on the principal ray and the same formulae can then be used after a suitable rotation of axes. However, Wynne suggests that in practical anamorphic systems there is no advantage to using crossed cylinders: spherical surfaces combined with parallel axis cylinders give adequate design freedom.

11.3 Aberrations of diffraction gratings

We noted in Chapter 5 the paradox that raytracing can be applied to diffractive optical systems; it follows from this that aberration theory can be developed for them also and in this section we sketch, again without detailed algebra, the theory for conventionally ruled gratings on plane and concave spherical surfaces. We follow the treatment in the review "Aberration Theory of Gratings and Grating Mountings" (W. T. Welford, *Progress in Optics*, **IV**, 241–280, 1965).

Figure 11.3 shows a concave grating with radius of curvature R and grating constant σ, as in Fig. 5.4. Consider two points A and A'; the condition that the ray from A to O, the centre of the grating, shall be diffracted to A' is stationarity of the optical path length for small displacements of the point of incidence from O, allowing for increments of a whole number of wavelengths as we cross a ruling (see Section 5.4). Thus we have to write down the difference between the optical path from A to A' via O and that from A to A' via another point P on the grating and expand it in powers of the coordinates (x, y) of P.

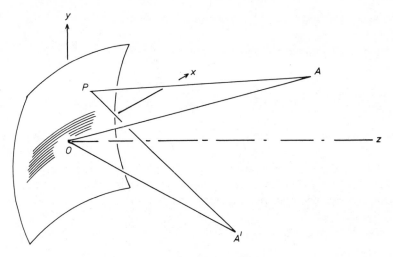

Fig. 11.3. A concave diffraction grating

The terms in different powers of x and y can then be picked out as different aberrations. Figure 11.4 is a view in the y–z section to show the notation— angles of incidence and diffraction, α, α'; distances OA, OA' and r, r'; x-coordinates of A and A', ξ, ξ' (not shown). The optical path length W from A to

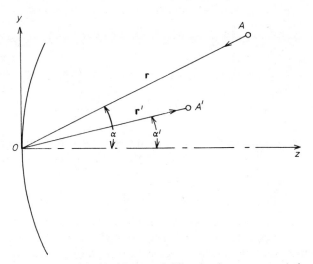

Fig. 11.4. Angles of incidence and diffraction for a concave grating

A' is easily found to be, as far as quadratic terms,

$$W = r + r' - (\sin \alpha + \sin \alpha')y + \frac{1}{2}\left\{\left(\frac{1}{r} - \frac{\cos \alpha}{R}\right) + \left(\frac{1}{r'} - \frac{\cos \alpha'}{R}\right)\right\}x^2$$

$$+ \frac{1}{2}\left\{\left(\frac{\cos^2 \alpha}{r} - \frac{\cos \alpha}{R}\right) + \left(\frac{\cos^2 \alpha'}{r'} - \frac{\cos \alpha'}{R}\right)\right\}y^2$$

$$+ \left\{\frac{\xi^2}{r^2}\sin \alpha + \frac{\xi'^2}{r'^2}\sin \alpha'\right\}y - \left(\frac{\xi}{r} + \frac{\xi'}{r'}\right)x + \cdots \tag{11.6}$$

As was shown in Section 5.4 we require W to be stationary for small displacements of P in the x direction and stationary apart from a linear term in y for displacements in the y direction, i.e. we want

$$\frac{\partial W}{\partial x} = 0, \qquad \frac{\partial W}{\partial y} = \frac{m\lambda}{\sigma} \tag{11.7}$$

where m is the order of diffraction. The first of eqns (11.7) gives

$$\frac{\xi}{r} + \frac{\xi'}{r'} = 0 \tag{11.8}$$

and the second

$$\sin \alpha + \sin \alpha' = \frac{m\lambda}{\sigma}. \tag{11.9}$$

This is known as the *grating equation* since it determines the directions of diffracted rays for a given direction of incidence. If we now put

$$S = \frac{1}{r} - \frac{\cos \alpha}{R}, \qquad T = \frac{\cos^2 \alpha}{r} - \frac{\cos \alpha}{R}, \qquad \text{etc.} \tag{11.10}$$

and apply eqns (11.8) and (11.9) we find that eqn (11.6) takes the form

$$W = \text{const.} + \frac{1}{2}(T + T')y^2 + \frac{1}{2}(S + S')x^2 + \frac{1}{2}\frac{m\lambda}{\sigma}\frac{\xi'^2}{r'^2}y + \cdots \tag{11.11}$$

We require a line focus parallel to the grating rulings in order to get spectroscopic resolution and we ensure this by putting the coefficient of y^2 equal to zero:

$$T + T' = 0. \tag{11.12}$$

This determines the locus of spectrum line foci. Of the remaining terms in eqn (11.11), that in x^2 corresponds to astigmatism and that in $\xi'^2 y$ to a curvature of the spectrum line, since it is a displacement of the image point in the y direction which is quadratic in ξ'.

In most arrangements of the concave grating the residual astigmatism is very large compared to that allowed in ordinary image-forming systems; this might be expected, since the grating is being used simultaneously for two very different purposes, spectral dispersion and image formation, and at the large angles of incidence and diffraction usuaily used we should expect large astigmatism in an image-forming system.

The higher order terms up to the fourth degree are examined in the reference given above; it is shown, for example, that types of coma and spherical aberration occur which are not present in axisymmetric systems.

Plane gratings are an important special case; they are generally used in a collimated beam with auxiliary collimating and focusing optics, usually mirrors. If in eqn (11.6) we put $\xi/r = \beta$, $\xi'/r' = \beta'$ and let $1/R$, $1/r$ and $1/r'$ tend to zero we obtain the properties of the plane grating in collimated light. Then with the obvious condition $\beta + \beta' = 0$ the only non-zero term in eqn (11.6) is

$$\tfrac{1}{2}(\sin \alpha + \sin \alpha')\beta'^2 y \tag{11.13}$$

which corresponds as we have already noted to a curvature of the spectrum line.

11.4   Aberrations of holographic optical elements

We recall from Section 5.6 that a holographic optical element may be formed on a non-plane substrate with aberrated forming beams and that the readout or reconstruction wavelength may be different from the recording wavelength; in fact this last is probably the most usual situation. The equivalent of primary aberration formulae for the most general geometry would be so involved as to be useless for practical purposes. We shall give only some special cases which illustrate the general method. First we give the primary aberration expressions for plane holograms with no aberrations in the forming beams and with axial symmetry (Section 11.4.1) and secondly we consider holograms on certain curved substrates which have special finite aberration properties (Sections 11.4.2 and 11.4.3).

11.4.1   Plane hologram optical elements

Figure 11.5 shows the geometry for recording a hologram in wavelength λ with recording beams focused at A_r and A_0. These foci are at conjugates l_r and l_0 and we take a typical point of incidence P with coordinate y from the origin O. We assume that the readout beam of wavelength λ' will be focused at A_r' and that we are looking for an image point at A_0', these two points being respectively at conjugates l_r' and l_0'. Let the point P correspond to a certain whole number of

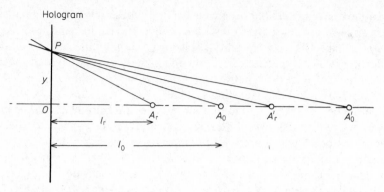

FIG. 11.5. Primary spherical aberration for a plane holographic optical element

fringes on the hologram from centre to edge. Then this number is

$$N = \frac{2\pi}{\lambda}\{OA_0 - PA_0 - (OA_r - PA_r)\}. \tag{11.14}$$

Then if A_0' is to be an mth order image we require that

$$\frac{2\pi}{m\lambda'}\{OA_0' - PA_0' - (OA_r' - PA_r')\} = N \tag{11.15}$$

and any deviations from this condition represent aberrations. The wavefront aberration is therefore given by

$$W = \lambda'\left[\frac{1}{m\lambda'}\{OA_0' - PA_0' - (OA_r' - PA_r')\}\right.$$

$$\left. - \frac{1}{\lambda}\{OA_0 - PA_0 - (OA_r - PA_r)\}\right]. \tag{11.16}$$

By Pythagoras' theorem we have, as far as the fourth degree terms in y,

$$PA_r = l_r\left\{1 + \frac{1}{2}\frac{y^2}{l_r^2} - \frac{1}{8}\frac{y^4}{l_r^4} + \cdots\right\}$$

and similarly for the other distances, so that

$$W = \lambda'\left[\frac{1}{2}y^2\left\{\frac{1}{m\lambda'}\left(\frac{1}{l_r'} - \frac{1}{l_0'}\right) - \frac{1}{\lambda}\left(\frac{1}{l_r} - \frac{1}{l_0}\right)\right\}\right.$$

$$\left. + \frac{1}{8}y^4\left\{\frac{1}{m\lambda'}\left(\frac{1}{l_0'^3} - \frac{1}{l_r'^3}\right) - \frac{1}{\lambda}\left(\frac{1}{l_0^3} - \frac{1}{l_r^3}\right)\right\}\right].$$

But the quadratic terms in y must vanish if A_0' is to be a Gaussian image point, so that for the primary spherical aberration term we have

$$W = \tfrac{1}{8} y^4 \left[\frac{1}{m} \left(\frac{1}{l_0'^3} - \frac{1}{l_r'^3} \right) - \frac{\lambda'}{\lambda} \left(\frac{1}{l_0^3} - \frac{1}{l_r^3} \right) \right].$$ (11.17)

If $m = 1$, $\lambda' = \lambda$ and $l_r' = l_r$, then eqn (11.17) shows that there is no primary spherical aberration for the image point at l_0', but, as is well known from the elementary theory of holography, under these conditions the image point is free from spherical aberration of all orders. The particular application of eqn (11.17) is that it shows how to change the readout coordinates l_r' and l_0' to eliminate the primary spherical aberration introduced when it is necessary to have λ' not equal to λ (or, what is perhaps more useful, how to choose l_r and l_0 so that there will be no aberration at given l_r' and l_0').

To examine the off-axis aberrations of a holographic optical element constructed as in Fig. 11.5 we refer to Fig. 11.6. Let the readout beam be

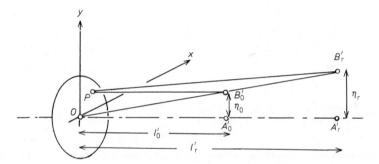

FIG. 11.6. Off-axis primary aberrations for a plane holographic optical element

focused at B_r' at a distance η_r from A_r'; we look for an image at B_0' distant η_0 from A_0' and to do this we examine the optical paths from points $P(x, y)$ and O on the hologram to B_r' and B_0'. The appropriate difference of optical path lengths expressed in units of phase is

$$\frac{2\pi}{m\lambda'} \cdot \{ OB_0' - PB_0' - (OB_r' - PB_r') \}$$

$$- \frac{2\pi}{\lambda} \{ OA_0 - PA_0 - (OA_r - PA_r) \}.$$ (11.18)

Again using Pythagoras' theorem and expanding as far as fourth powers in x

and y we find

$$
\mathrm{OB'_r} - \mathrm{PB'_r} = -\frac{(x^2 + y^2 - 2\eta_r y)}{2l'_r} + \frac{1}{8l'^3_r}(x^2 + y^2)^2
$$

$$
- \frac{1}{2l'^3_r}\eta_r y(x^2 + y^2) + \frac{1}{2l'^3_r}\eta_r^2 y^2
$$

$$
+ \frac{1}{4l'^3_r}\eta_r^2(x^2 + y^2) - \frac{1}{2l'^3_r}\eta_r^3 y \qquad (11.19)
$$

with a similar expression for $\mathrm{OB'_0} - \mathrm{PB'_0}$. The linear and quadratic terms must vanish if $\mathrm{B'_0}$ is to be the Gaussian image point, so that $\eta_r/l'_r = \eta_0/l'_0$. Then omitting the spherical aberration term we are left with the following expression for the off-axis primary aberration:

$$
-\frac{1}{2l'_0}\left(\frac{1}{l'^2_0} - \frac{1}{l'^2_r}\right)\eta_0 y(x^2 + y^2) + \frac{3}{4l'^2_0}\left(\frac{1}{l'_0} - \frac{1}{l'_r}\right)\eta_0^2 y^2
$$

$$
+ \frac{1}{4l'^2_0}\left(\frac{1}{l'_0} - \frac{1}{l'_r}\right)\eta_0^2 x^2. \qquad (11.20)
$$

The terms in this expression correspond respectively to coma and to tangential and sagittal field curvature; since we have assumed the stop to be at the hologram there is no distortion. As might be expected, the recording geometry does not enter into eqn (11.20). We note that there is no term corresponding to the Petzval sum and we see that the off-axis aberrations vanish simultaneously if and only if $l'_0 = l'_r$. Clearly coma would set a limit on the performance of such an element, an inescapable limitation for single holographic elements on plane substrates. Thus to get a corrected field of view with a large convergence angle an axially symmetric plane hologram must be combined with another or with reflecting or refracting components.

11.4.2   Hologram optical elements on curved substrates

We consider only substrates which have revolution symmetry and in this section we calculate astigmatic effects along a principal ray, i.e. we seek the analogues of the formulae of Thomas Young (Section 9.8). Figure 11.7 shows a section of the hologram in a plane through the axis of symmetry; the local principal curvatures at the point of incidence O are c_t and c_s. The principal rays of the recording beam are $\mathrm{OA_r}$ and $\mathrm{OA_0}$; the readout and image beams are not shown but the same notation is used as in Section 11.4.1. If we set up a local coordinate system as in the figure the equation of the surface as far as

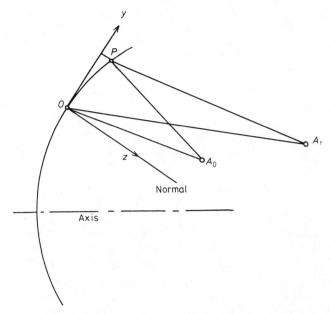

FIG. 11.7. Astigmatism produced by a holographic optical element on a non-plane substrate

quadratic terms is

$$z = \tfrac{1}{2}(c_s x^2 + c_t y^2) + \cdots \tag{11.21}$$

The condition for imaging in the y–z section is as before

$$\frac{1}{m\lambda'}\{OA_0' - PA_0' - (OA_r' - PA_r')\}$$

$$-\frac{1}{\lambda}\{OA_0 - PA_0 - (OA_r - PA_r)\}. \tag{11.22}$$

Expanding $OA_r - PA_r$ as far as quadratic terms we obtain

$$OA_r - PA_r = y\sin I_r + \tfrac{1}{2}y^2 c_t \cos I_r - \frac{\tfrac{1}{2}y^2\cos^2 I_r}{OA_r} + \cdots \tag{11.23}$$

with similar expressions for the other terms in eqn (11.22). Substituting all these in eqn (11.22) we see that when the coefficient of y is equated to zero we recover the raytracing formula, eqn (5.33). Then the quadratic terms give the position of the focus A'_0; putting $OA_r = t_r$ etc., we find

$$\frac{1}{m\lambda'}\left\{c_t(\cos I'_0 - \cos I'_r) - \frac{\cos^2 I'_0}{t'_0} + \frac{\cos^2 I'_r}{t'_r}\right\}$$

$$- \frac{1}{\lambda}\left\{c_t(\cos I_0 - \cos I_r) - \frac{\cos^2 I_0}{t_0} + \frac{\cos^2 I_r}{t_r}\right\} = 0. \qquad (11.24)$$

In the x–z section a similar calculation yields

$$\frac{1}{m\lambda'}\left\{c_s(\cos I'_0 - \cos I'_r) - \frac{1}{s'_0} + \frac{1}{s'_r}\right\}$$

$$- \frac{1}{\lambda}\left\{c_s(\cos I_0 - \cos I_r) - \frac{1}{s_0} + \frac{1}{s_r}\right\} = 0. \qquad (11.25)$$

The correspondence with the results of Section 9.8 is obvious. Equations (11.24) and (11.25) were first given by R. W. Smith (*Optics Communications*, **21**, 106–109, 1977). The same author showed also that they lead to a remarkable set of special cases of hologram optical elements which are free from astigmatism (*Optics Communications*, **19**, 245–247, 1976 and **21**, 102–105, 1977). These occur when the hologram is formed on a quadric of revolution and the principal rays pass through the two foci of the quadric. For, put

$$\cos I_0 = \cos I_r = \cos I'_0 = \cos I'_r$$

in eqns (11.24) and (11.25) and recall from eqn (9.74) that

$$c_t = c_s \cos^2 I$$

for a ray from one focus to the other. Then it follows that there is no astigmatism for readout and image points along any principal ray. Figure 11.8 summarizes some typical geometries. The special case when the hologram is formed on a plane is also shown, since this satisfies the two conditions above although the foci are in this case both at infinity.

11.4.3   Aplanatic holographic optical elements

Our final special case is that of a hologram on a curved substrate which is aplanatic—the analogue of the components described in Section 8.8. We first recall from Chapter 9 that the Abbe sine condition for aplanatism becomes for an object at infinity

$$y/h = \sin U'/\sin u$$

so that if the optical system is "thin" it must look as in Fig. 11.9 (this is sometimes expressed by the statement that the second principal "plane" must

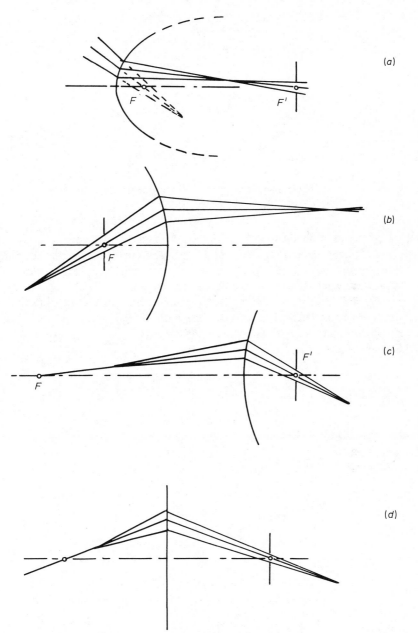

FIG. 11.8. Examples of stigmatic imagery by narrow pencils at holographic optical elements. (a) Ellipsoid, (b) paraboloid, (c) hyperboloid and (d) plane. In the first three cases the pupils are at the foci of the quadrics and in the last case the pupil magnification is −1

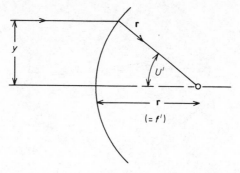

FIG. 11.9. Aplanatism for one infinite conjugate with a holographic optical element on a spherical substrate

be a spherical surface centred on the axial image point). Thus a hologram optical element formed on a surface of radius of curvature r with recording beams respectively parallel and normally incident will be aplanatic if used in the same wavelength and with a collimated incident beam.

To generalize this consider a hologram on a spherical surface of radius of curvature r recorded with beams converging to foci at distances l and l' and read out with a beam of the same wavelength converging to l, as in Fig. 11.10. The sine condition requires

$$\sin U'/\sin U = \text{const.} = l/l'. \tag{11.26}$$

From the diagram,

$$\sin^2 U = \frac{r^2 - (r - z)^2}{(l - z)^2 + r^2 - (r - z)^2}$$

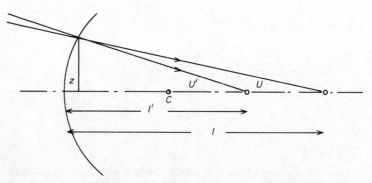

FIG. 11.10. Calculating the aplanatic conjugates for a hologram optical element on a spherical substrate

and substituting this and a similar expression for $\sin^2 U'$ into eqn (11.26) we obtain after some reduction the condition

$$\frac{1}{r}\left(\frac{1}{l'} - \frac{1}{l}\right) = \frac{1}{l'^2} - \frac{1}{l^2}. \tag{11.27}$$

Thus the requirement for aplanatism is either

$$l' = l \tag{11.28}$$

or

$$\frac{1}{l'} + \frac{1}{l} = \frac{1}{r}. \tag{11.29}$$

Equation (11.28) is trivial, since it is equivalent to normal incidence at a spherical reflecting or refracting surface, but eqn (11.29) is a non-trivial result giving a range of geometries for which aplanatism can be obtained. For a discussion of the effects of using different recording and readout wavelengths see W. T. Welford, *Optics Communications*, **15**, 46–49 (1975).

12 Thin Lens Aberrations

12.1 The thin lens variables

The notions of a thin lens and of combinations of thin lenses are essential in setting out optical systems on the basis of their Gaussian properties, conjugate positions, pupil positions, magnification, relative aperture, etc. Equally the thin lens concept is useful in the early stages of aberrational design, i.e. where the approximate powers, bendings and positions of lens elements are being determined for the required aberration correction. Relatively simple formulae are available for the primary aberrations, (including C_I and C_{II}), and these can very often enable the designer to see if a proposed layout offers a reasonable chance of controlling the primary aberrations. In some cases the exact solutions are available if suitable non-linear simultaneous equations are solved; for example a thin lens solution for the celebrated Cooke triplet objective invented by H. Dennis Taylor can be formulated in this way. The thin lens aberration formulae also lead to several useful general rules about the kinds of correction possible in certain cases; for example in Section 12.2 we shall see that a system of thin lenses in contact cannot have zero astigmatism unless the total power is zero.

In deriving the Seidel sums in Chapter 8 we added terms surface by surface and these involved the parameters of the optical system (curvatures, refractive indices, etc.) and the object–image conjugates. These were specified by parameters and variables which appeared naturally in the derivations as the formulae were tidied up and put into convenient forms for computing. We now have to choose corresponding parameters and variables to be used in thin lens theory. In this development we shall follow mainly an exposition by C. G.

Wynne[†] in which many useful results are compactly collected. We shall, following Wynne, generalize the notion of a thin lens slightly to mean a thin component of refractive index $n_0 n$ in a medium of index n_0; this enables us to include the case of a thin air gap with curved surfaces between glasses of equal indices, a situation which sometimes occurs. Recalling the definition of the power of a system given in Section 3.2 we have for the power of this generalized thin lens

$$K = n_0(n - 1)(c_1 - c_2). \tag{12.1}$$

Next we need to specify the shape or bending of the lens (Fig. 3.19) and the conjugates. Here there are many possibilities but the choice which leads to formulae which display most clearly the aberrational properties of thin lenses goes back to Henry Coddington;[‡] he used symmetrical variables which, apart from trivial factors, are defined as follows:

Shape or bending variable:

$$B = \frac{n_0(n - 1)(c_1 + c_2)}{K} = \frac{c_1 + c_2}{c_1 - c_2}$$

Conjugate or magnification variable:

$$C = \frac{n_0(u_1 + u'_2)}{hK} = \frac{u_1 + u'_2}{u_1 - u'_2}$$

$$\left.\vphantom{\begin{array}{c}a\\b\\c\\d\end{array}}\right\} \tag{12.2}$$

These are symmetrical in the curvatures and conjugates respectively and they are dimensionless. Also they are normalized (division by K), so that the overall scale is eliminated. Thus we find that for a lens of positive power, $B = 0$ makes it equiconvex, $B = 1$ makes it planoconvex with the convex side to the left and $B = -1$ makes it planoconvex with the plane side to the left. Similarly $C = 0$ means equal conjugates, i.e. $l = -l'$, $C = 1$ means the object is at the first principal focus and $C = -1$ means the object is at infinity. The following relations between the Coddington variables and the convergence angles, etc. will be useful in Section 12.2.

$$c_1 = \frac{K}{2n_0(n - 1)}(B + 1)$$

$$c_2 = \frac{K}{2n_0(n - 1)}(B - 1)$$

$$u_1 = \frac{hK}{2n_0}(C + 1)$$

$$u'_2 = \frac{hK}{2n_0}(C - 1)$$

$$\left.\vphantom{\begin{array}{c}a\\b\\c\\d\\e\\f\\g\\h\end{array}}\right\} \tag{12.3}$$

† *Optica Acta* **8**, 255–265 (1961).
‡ "A treatise on the reflexion and refraction of light", London (1829).

Also for the transverse magnification m we have

$$m = \frac{u_1}{u_2'} = \frac{C + 1}{C - 1}. \tag{12.4}$$

Many different sets of variables for thin lens theory have been used, some unsymmetrical and others symmetrical and more or less resembling the original Coddington variables as given above. For practical numerical work it is possible to make a good case for other choices, in particular for variables which are not dimensionless and normalized; however, our aim in this chapter is not so much to give the best system for practical computation but rather to display as simply and clearly as possible the main aberrational properties of thin lens systems. For this purpose the present choice seems best.

12.2 Primary aberrations of a thin lens with the pupil at the lens

We can calculate the Seidel sums for a thin lens, i.e. a pair of surfaces, by taking a symbolic raytrace through the lens, as was done in Section 3.8, substituting in eqns (8.41) and then replacing u, c, A, etc. by their equivalents in the thin lens variables as obtained from eqns (12.3). To simplify the algebra it is best to consider first the case when the aperture stop is at the lens, i.e. at both surfaces since it is thin, and then to allow for eccentric pupils by means of the stop-shift formulae of Chapter 8. The calculation is lengthy but not particularly difficult and we give here a brief outline.

The powers of the individual surfaces are $n_0(n - 1)c_1$ and $-n_0(n - 1)c_2$; the raytrace gives, for example, for the convergence angle inside the lens $\{u_1 - (n - 1)c_1h\}/n$ and for the final convergence angle $u_2' = u_1 - hK/n_0$. Thus we obtain:

$$-S_1 = n_0(hc_1 + u_1)^2h\left\{\frac{u_1 - (n - 1)hc_1}{n^2} - u_1\right\}$$

$$+ n_0(hc_2 + u_2')^2h\left\{u_2' - \frac{u_1 - (n - 1)hc_1}{n^2}\right\}.$$

We next substitute from eqns (12.3) and it follows in the customary "few lines" that

$$S_1 = \frac{h^4K^3}{4n_0^2}\left\{\left(\frac{n}{n - 1}\right)^2 + \frac{n + 2}{n(n - 1)^2}\left(B + \frac{2(n^2 - 1)}{n + 2}C\right)^2 - \frac{n}{n + 2}C^2\right\}. \tag{12.5}$$

Similarly we obtain for coma and astigmatism with central stop,

$$S_{\mathrm{II}} = -\frac{h^2 K^2 H}{2n_0^2} \left\{ \frac{n+1}{n(n-1)} B + \frac{(2n+1)}{n} C \right\}, \tag{12.6}$$

$$S_{\mathrm{III}} = \frac{H^2 K}{n_0^2}. \tag{12.7}$$

The Petzval sum is, as we already know, independent of the conjugates and the stop position; we have

$$S_{\mathrm{IV}} = \frac{H^2 K}{n_0^2 n}. \tag{12.8}$$

If the principal ray meets the thin lens at the axis it is transmitted undeviated, so the distortion vanishes:

$$S_{\mathrm{V}} \equiv 0 \tag{12.9}$$

For longitudinal colour we obtain either by working through from eqn (10.19) or more directly from eqn (10.2),

$$C_{\mathrm{I}} = \frac{h^2 \, K \, \delta n}{n-1} = \frac{h^2 K}{V}. \tag{12.10}$$

Finally, the transverse colour must be zero by the same argument as for distortion:

$$C_{\mathrm{II}} \equiv 0. \tag{12.11}$$

We can immediately make several useful deductions from these results. We already know that spherical aberration varies as the fourth power of the aperture; from eqn (12.5) we see that if the other variables are fixed S_{I} depends on the cube of the power K. Likewise coma varies quadratically and astigmatism linearly as K. From eqn (12.7) the astigmatism of a combination of thin lenses in contact, with the stop at the lenses, is proportional to the sum of their powers and since the total power is equal to the sum of the individual powers we see that no thin system can have zero astigmatism unless it has zero power, the result mentioned in Section 12.1. The same is not strictly true for the Petzval sum (eqn (12.8)) since the factor $1/n$ can vary for the different components, but in practice the available range is too small to be useful and it is practically true that a thin system cannot have zero Petzval sum if its power is non-zero.

We now return to eqn (12.5) and consider S_{I} in more detail. If we keep the conjugates the power fixed and vary the shape B, i.e. bend the lens, it can be seen that S_I depends quadratically on the bending. We can plot $4n_0^2 S_I/h^4 K^3$ as a function of B for given n and C (conjugate variable) and obtain a series of parabolas; all parabolas for given n have the same curvature at the vertex, $2(n + 2)/n(n - 1)^2$.

The vertex of each parabola, i.e. the minimum of S_I, occurs at

$$B = -\frac{2(n^2 - 1)}{n + 2} C \tag{12.12}$$

and the value of the minimum is

$$S_{I_{min}} = \frac{h^4 K^3}{4n_0^2}\left\{\left(\frac{n}{n - 1}\right)^2 - \frac{n}{n + 2} C^2\right\}. \tag{12.13}$$

From eqns (12.12) and (12.13) the locus of the minima is another parabola,

$$S_{I_{min}} = \frac{h^4 K^3}{4n_0^2}\left\{\left(\frac{n}{n - 1}\right)^2 - \frac{n(n + 2)}{4(n^2 - 1)^2} B^2\right\}. \tag{12.14}$$

Figure 12.1 shows the parabolas plotted with $C = 0$ for values of n of 1·5, 1·7, 1·9, 2·4 and 4·0; also the parabola represented by eqn (12.14), which represents the locus of minima as C varies, is plotted in broken line for $n = 1·5$, and values of C equal to 1, 2 and 3 are marked on it. Thus to get the parabola for S_I for $n = 1·5$ and $C = 2$, say, the parabola shown for $n = 1·5$ is translated without rotation until its vertex lies on the point marked 2 on the broken line parabola.

It can be seen that the spherical aberration of a lens of positive power cannot be made negative except for numerically large values of C, i.e. there must be one virtual conjugate; at the same time numerically large values of B are required, so the lens has to have a deep meniscus shape. From eqn (12.13) the value of C required for zero S_I is given by

$$C^2 = \frac{n(n + 2)}{(n - 1)^2}, \tag{12.15}$$

i.e. a value of $|C|$ between 4·5 and 2·5. From eqn (12.14) the value of B needed to go with this is given by

$$B^2 = \frac{4n(n + 1)^2}{n + 2}, \tag{12.16}$$

i.e. between 3·3 and 8.

Figure 12.1 illustrates very clearly a general rule in optical design that, other things being equal, higher refractive indices lead to smaller aberrations. The effect is very marked in the infrared region, where materials with $n \sim 4$ are

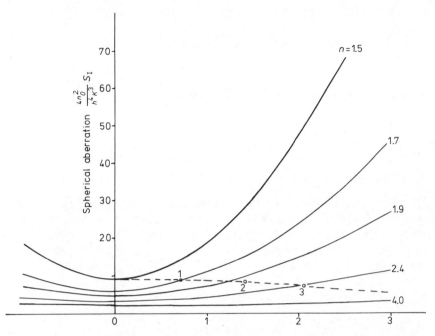

FIG. 12.1. Primary spherical aberration of thin lenses as a function of the shape variable B. The full lines are for conjugate variable $C = 0$ and for the refractive indices indicated. For other values of C the curves are translated as indicated in the text; for example, the curve for $n = 1.5$ is moved so that its vertex stays on the broken line, the appropriate values of C being as labelled

available. In the visible, optical glasses with a range of dispersions and with refractive indices in the order of 1·8 have become available only since about 1950 and these have had considerable effect in improving designs. A minor further advantage of such glasses is that they can be anti-reflection coated quite well with a single layer of magnesium fluoride; $n = 1.9$ gives the theoretically perfect match but any index higher than 1·75 is close enough to make it hardly worth bothering with a multilayer anti-reflection coating unless very wide-band performance is wanted.

Equation (12.12) gives the best shape lens for given conjugates; thus for equal conjugates $(C = 0)$ the best shape is, not surprisingly, $B = 0$, or equiconvex. For $C = \pm 1$, object or image at infinity, the best shape is approximately planoconvex with the convex side towards the infinity conjugate; however, the exact shape is usually slightly different, depending on the refractive index; $n = 1.69$ gives exactly a planoconvex shape for minimum aberration.

Equation (12.6) shows that S_{II} is a linear function of bending when the stop is at the lens; thus the shape for zero linear coma is given by

$$B = -\frac{(n-1)(2n+1)}{n+1}C;$$

numerically this is close to the shape for minimum spherical aberration, particularly for the higher values of refractive index.

12.3   Primary aberrations of a thin lens with remote stop

Thin lens aberration formulae for a remote pupil are, of course, essential for design work involving off-axis aberrations. We can get these formulae very easily by following the method used in Section 8.5 for stop-shift effects. We can define an eccentricity parameter E for a thin lens in exactly the same way as for a surface: HE is equal to $\bar{h}/h$ at the lens and it is computed for a thin lens system in the same way, working outwards from the stop. We then have for the collected Seidel aberrations of a thin lens

$$S_{\mathrm{I}} = \frac{h^4 K^3}{4n_0^2}\left\{\left(\frac{n}{n-1}\right)^2 + \frac{n+2}{n(n-1)^2}\left(B + \frac{2(n^2-1)}{n+2}C\right)^2 - \frac{n}{n+2}C^2\right\}$$

$$S_{\mathrm{II}} = -\frac{h^2 K^2 H}{2n_0^2}\left\{\frac{n+1}{n(n-1)}B + \frac{(2n+1)}{n}C\right\} + HE.S_{\mathrm{I}}$$

$$S_{\mathrm{III}} = \frac{H^2 K}{n_0^2} - HE\cdot\frac{h^2 K^2 H}{n_0^2}\left\{\frac{n+1}{n(n-1)}B + \frac{(2n+1)}{n}C\right\} + (HE)^2 S_{\mathrm{I}}$$

$$S_{\mathrm{IV}} = \frac{H^2 K}{n_0^2 n}.$$

$$S_{\mathrm{V}} = HE\cdot\frac{H^2 K}{n_0^2}\left\{3 + \frac{1}{n}\right\} - 3(HE)^2\frac{h^2 K^2 H}{2n_0^2}\left\{\frac{n+1}{n(n-1)}B + \frac{(2n+1)}{n}C\right\}$$
$$+ (HE)^3 S_{\mathrm{I}}$$

$$C_{\mathrm{I}} = \frac{h^2 K}{V}$$

$$C_{\mathrm{II}} = HE\cdot C_{\mathrm{I}}$$

$$\tag{12.17}$$

The monochromatic Seidel sums considered first as functions of B, the shape variable, all have very similar quadratic forms, apart from S_{IV}, since they all contain S_{I}. If we plot S_{I}, S_{II}, S_{III} and S_{V} as functions of B we get a set of parabolas with vertical axes. The vertex curvatures are as follows:

$$
\left.\begin{array}{ll}
S_{\mathrm{I}}: & \dfrac{n+2}{n(n-1)^2}\cdot\dfrac{h^4 K^3}{2n_0^2} \\[4mm]
S_{\mathrm{II}}: & \dfrac{HE(n+2)}{n(n-1)^2}\cdot\dfrac{h^4 K^3}{2n_0^2} \\[4mm]
S_{\mathrm{III}}: & (HE)^2\,\dfrac{n+2}{n(n-1)^2}\cdot\dfrac{h^4 K^3}{2n_0^2} \\[4mm]
S_{\mathrm{V}}: & (HE)^3\,\dfrac{(n+2)}{n(n-1)^2}\cdot\dfrac{h^4 K^3}{2n_0}
\end{array}\right\}.
\tag{12.18}
$$

The abscissae of the vertices, i.e. of the axes of the parabolas, are

$$
\left.\begin{array}{ll}
S_{\mathrm{I}}: & -\dfrac{2(n^2-1)}{n+2}\,C \\[4mm]
S_{\mathrm{II}}: & \dfrac{(n^2-1)}{n+2}\left(-2C+\dfrac{1}{Eh^2 K}\right) \\[4mm]
S_{\mathrm{III}}: & \dfrac{(n^2-1)}{(n+2)}\left(-2C+\dfrac{2}{Eh^2 K}\right) \\[4mm]
S_{\mathrm{V}}: & \dfrac{(n^2-1)}{n+2}\left(-2C+\dfrac{3}{Eh^2 K}\right)
\end{array}\right\}.
\tag{12.19}
$$

Finally the ordinates of the vertices, i.e. the stationary values of the aberrations with respect to varying shape, are

$$
S_{\mathrm{I}_{\text{stat.}}} = \frac{h^4 K^3}{4n_0^2}\left\{\left(\frac{n}{n-1}\right)^2 - \frac{n}{n+2}C^2\right\}
$$

$$
S_{\mathrm{II}_{\text{stat.}}} = HE.S_{\mathrm{I}_{\text{stat.}}} - \frac{1}{2(n+2)}\cdot\frac{h^2 K^2 HC}{n_0^2} - \frac{(n+1)^2}{n(n+2)}\cdot\frac{HK}{2n_0^2 E}
$$

$$
S_{\mathrm{III}_{\text{stat.}}} = (HE)^2.S_{\mathrm{I}_{\text{stat.}}} - \frac{1}{n+2}\cdot\frac{h^2 K^2 H^2 EC}{n_0^2} - \frac{(n^2+2n+2)H^2 K}{n(n+2)}\cdot\frac{1}{n_0^2}
$$

$$
S_{\mathrm{V}_{\text{stat.}}} = (HE)^3.S_{\mathrm{I}_{\text{stat.}}} - \frac{3}{2(n+2)}\cdot\frac{h^2 K^2 H^3 E^2 C}{n_0^2} - \frac{(3n^2+4n+5)}{2n(n+2)}\cdot\frac{H^2 K}{n_0^2}.
$$

$$\tag{12.20}$$

From eqn (12.18) the curvatures are successively multiplied by HE and from eqn (12.19) the axes are successively shifted by $(n^2 - 1)/(n + 2)Eh^2K$. The vertex ordinates do not have a regular progression unless the stop is very remote, i.e. very large HE, when it can be seen that the first terms in eqn (12.20) are dominant and successive values are multiplied by the factor HE.

These results cannot, of course, be used in the limiting case when E tends to zero, i.e. the stop is at the lens, since then coma becomes a linear function of bending, astigmatism is constant and distortion is zero, as in eqns (12.5) to (12.9). Another special case is a field lens (Section 3.9), i.e. when an intermediate image is at the lens; the conjugate variable C then tends to infinity and since h, the incidence height, tends to zero it can be seen that some indeterminacies occur in the expressions for the Seidel sums. It is not difficult to resolve these indeterminacies by examining the formulae or alternatively we can return to the surface-by-surface forms of the Seidel sums given in Chapter 8. Either way it is easily shown that all Seidel aberrations except S_{IV} vanish and likewise C_I and C_{II} vanish. The value of S_{IV} is given simply by eqn (12.8). Thus a thin field lens introduces no aberrations but field curvature.

It is sometimes useful to consider a thin lens as being aspherized; an important example is a zero power system or *aspheric plate*. It is easily seen that this can be incorporated into the thin lens scheme very easily by determining its contribution to S_I in the usual way and then adding the contributions to the other aberrations if the stop is not at the aspheric plate by means of the formulae in Section 8.6.

12.4   Aberrations of plane parallel plates

Plane parallel plates in the form of windows or prisms occur often in optical systems and the Seidel sums take particularly simple forms. It is appropriate to include these results in a chapter on thin lenses because it is often desirable to include a prism in a design at the thin lens layout stage. Also it is of interest that H. Dennis Taylor[†] used a method of calculating the Seidel aberrations of a system with components of finite thicknesses by treating it as composed of thin lenses and plane parallel plates.

We suppose the plate has thickness d and refractive index n_0n and also it is surrounded by a medium of index n_0. Let the incident axial pencil have convergence angle u and incidence height h on the first surface, as in Fig. 12.2.

† "A System of Applied Optics", Macmillan, London (1906).

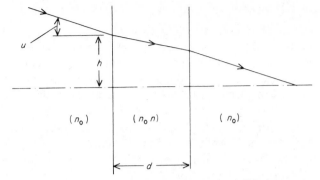

FIG. 12.2. Seidel aberrations of a plane-parallel plate

Then by applying eqns (8.41) it is easily shown that

$$S_I = -\frac{n_0(n^2 - 1)}{n^3} u^4 d$$

$$S_{II} = \frac{\bar{u}}{u} S_I$$

$$S_{III} = \left(\frac{\bar{u}}{u}\right)^2 S_I$$

$$S_{IV} = 0$$

$$S_V = \left(\frac{\bar{u}}{u}\right)^3 S_I$$

$$C_I = -n_0 \frac{\delta n}{n^2} u^2 d$$

$$C_{II} = -n_0 \frac{\delta n}{n^2} u\bar{u}d$$

$$(12.21)$$

where $\bar{u}$ is the convergence angle of the principal ray. It is easy to make a substitution for $\bar{u}$ in terms of the Lagrange invariant but this then brings in incidence heights explicitly; the forms given in eqns (12.21) show that the Seidel aberrations introduced by a plane parallel plate are unchanged by moving the plate along the optical axis, i.e. they are actually independent of incidence heights. The same is in fact true for total aberrations of finite rays. It can be seen also that a plane parallel plate introduces no aberrations if it is in a star space and that the magnitudes of all the aberrations vary linearly with the plate thickness.

12.5 Some examples

In this section we show how thin lens theory can be used by means of a number of examples. These examples do not of course cover all possibilities and the discussion is only taken far enough to indicate the general procedure.

In Chapter 10 we saw how to obtain an achromatic doublet from two thin lenses in contact. If adjacent surfaces have the same curvature they can be cemented in contact and we have then a single degree of freedom, the bending of the doublet as a whole, as in Fig. 12.3. We can plot the spherical aberration and coma of this system with, say the object at infinity and the stop at the lens and we shall, from Section 12.3, get a parabola and a straight line, as in Fig.

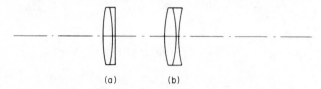

(a)   (b)

FIG. 12.3. Bending a doublet objective; in going from (a) to (b) each surface has the same increment of curvature

12.4, as a result of adding the two parabolas and straight lines for the individual components. If, as would normally be the case, the negative component has a sufficiently higher index than the positive and if the contact surface has a large enough curvature the S_I parabola will cut the horizontal axis and there will be two solutions for zero spherical aberration. In general the S_{II} line will not cut the axis at either point and so the doublet is not aplanatic. If the order of the glasses is reversed—"flint in front"—very similar graphs are obtained and again there are two solutions. The doublet is free from transverse colour for all positions of the stop.

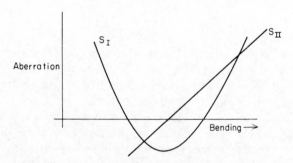

FIG. 12.4. Spherical aberration and coma as a function of bending for a doublet objective

We can get an aplanatic thin lens doublet in either of two ways. First we can choose the glasses in the cemented doublet carefully so that the S_{II} line crosses the axis at one of the crossing points of the parabola. The second possibility is to break the contact and bend the two components separately; clearly there are then enough degrees of freedom available to control S_I and S_{II} independently whatever the choice of glasses.

Similar reasoning would apply to the design of a doublet for any other conjugates.

If the stop is remote from the lens we can control S_{II} by taking a non-zero S_I and making use of the stop-shift equations; this does not necessarily mean that the system would have an objectionable amount of spherical aberration, for if HE had a fairly large value, say, 3, at the lens it could happen that S_I was below tolerance but $HE.S_I$ was large enough to control S_{II} usefully. To make use of this notion properly we need the concept of aberration tolerances, discussed in Chapter 13. Similarly for a remote stop non-zero values of S_I and S_{II} can be used to control S_{III}. However, these ideas are only of practical use for systems like microscope eyepieces which have very small apertures.

For a system of large aperture it is clear that S_I, S_{II} and S_{III} cannot all be made to vanish unless either it contains thick lenses or it consists of separated thin lenses. The classical example is the Cooke triplet photographic

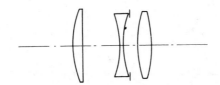

FIG. 12.5. A Cooke triplet photographic objective

objective, shown in Fig. 12.5; this contains precisely enough parameters to enable all primary aberrations to be controlled, provided the glass types are suitably chosen. The parameters may be divided into two groups, first the three powers and the two separations and secondly the three bendings; four parameters of the first group control the overall power, the two chromatic aberrations and the Petzval sum; the remaining parameter of the first group (the choice is not critical) and the three bendings together control S_I, S_{II}, S_{III} and S_V. The actual solutions have been obtained by a variety of methods of successive approximation. This simple design gives good correction as a photographic objective of about 50 mm focal length up to relative aperture $f/3\cdot5$ and field angle $\pm20°$; beyond that more complex designs are needed to control higher order aberrations.

If two doublets have powers K and $-K$ and they are spaced d apart the combination will have the non-zero power K^2d; if both are separately aplanatic the combination will have zero S_{III} and, if the glass pairs are the same in both doublets, zero S_{IV}. The result with an infinite conjugate on the negative side is an inverted telephoto system, as in Fig. 12.6, i.e. a system in

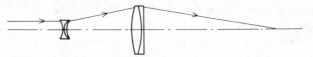

FIG. 12.6. An inverted telephoto system

which the efl is much less than the bfl; this is sometimes useful when access is limited behind the lens.

A field-flattening lens is a lens at or near a field surface of which the object is to correct field curvature; as we saw in Section 12.3 it cannot affect other aberrations. Figure 12.7 shows the principle as used in large aperture photographic objectives.

The Schmidt plate is the name given to the crucial element in the Schmidt camera (Fig. 12.8); basically the Schmidt camera is a concave spherical mirror with the aperture stop at its centre of curvature; under these conditions it is

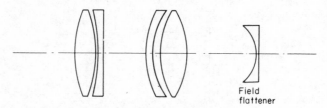

Field
flattener

FIG. 12.7. A high aperture objective of which the last component is a field flattener

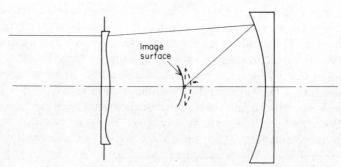

Image
surface

FIG. 12.8. The Schmidt camera; the aspheric plate is at the centre of curvature of the mirror and it forms the aperture stop

easy to see from symmetry that only spherical aberration and field curvature occur. The Schmidt plate is a thin aspheric plate at the stop; the aspherizing is chosen to correct the spherical aberration and in order to improve the higher order aberration correction a small Gaussian power is also given to the plate. Occasionally a (*convex*) field flattening lens is used as indicated in broken line.

Further examples on these lines can be found in the paper by C. G. Wynne quoted in Section 12.1 and in books on optical design.

13   Optical Tolerances

13.1   Design aberrations and manufacturing aberrations

It is a matter of common experience with optical designers that optical systems cannot be designed to be aberration-free for a single magnification except in certain trivial cases; even if we demand only freedom from point-imaging aberrations and in monochromatic light, the problem is insoluble over an extended field. The Schmidt camera is probably the closest approach to such a system, but it has higher order astigmatism because the aspheric plate is traversed obliquely by the off-axis pencils. Thus it is necessary to have optical tolerances to be applied to designs so that a design may be said to be good enough according to the tolerance system appropriate to the intended application. Such tolerances take the form of limits on the amounts of aberrations, individually or in stated combinations, which can be permitted.

The subject of tolerances on "manufacturing aberrations" is closely linked with that of design tolerances. When a designed system has been constructed it never conforms exactly to the computed design: curvatures, refractive indices, glass thicknesses and air spaces may be slightly wrong,† but also the surfaces may not be truly spherical, the glass may not be perfectly homogeneous and the components may not be mounted quite coaxially. The resulting effects, known as manufacturing aberrations, should likewise be subject to tolerance conditions. Some of the discussion in this chapter is relevant

† But in the most exacting work, it is usual to work testplates, measure their radii and then readjust the design by changing thicknesses and air spaces appropriately. Also the actual glass melt indices are used after catalogue values have been used for the first design.

to such effects, but in general it is impracticable to develop systematic tolerances for, say, decentring and manufacturers have to be guided by experience.

13.2   Some systems of tolerances

Optical tolerances can be discussed in several different ways. On the most elementary level we can take a geometrical optics approach and consider the spread of the rays in the image of a point as indicating the degradation of the image by aberrations. Thus the spot diagram or the ray intersection density (Section 7.7) would be compared with the scale of structure to be imaged. The geometrical optics approach is appropriate mainly to systems of moderate performance and to the earlier stages of design of higher quality systems. If a pencil from a point object were aberration-free, then geometrical optics would suggest the system was capable of infinite resolving power, but it is well known from physical optics that diffraction effects then limit the performance and the aberration-free point image is actually a diffraction pattern of finite width. More sophisticated systems are therefore based on the effect of aberrations on the physical optics point image and these are applicable to systems which are almost aberration-free.

Sometimes it is more appropriate to consider extended objects rather than single points; these extended objects may be single lines, edges or periodic structures; in each case the question of the nature of the illumination, i.e. whether it is coherent or incoherent, has to be considered. It can be seen from the above that the subject of optical tolerances is very extensive; it would be inappropriate to treat it in detail in a book which is mainly concerned with geometrical optics, but on the other hand, some discussion seems essential. In what follows, we give an outline of the principal ideas and methods, but we omit mathematical proofs.

13.3   Tolerances for diffraction-limited systems

If an optical system has no aberrations the image of a bright point object against a dark background, the so-called point spread function, is a rather complex diffraction figure of which the form depends on the shape of the exit pupil of the system. It is customary to make certain approximations in calculating the point spread function; these are that polarization effects are ignored, that convergence angles are small and that the exit pupil is large in diameter compared to the wavelength; these are usually known as the scalar diffraction theory approximations and it is an experimental fact that the scalar theory is very accurate under the conditions which usually hold in optics. If we also assume that the light transmission of the optical system is the same all over the pupil, i.e. that there is no varying absorption across the aperture,

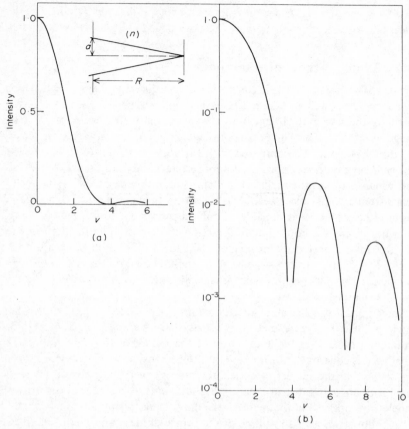

FIG. 13.1. The Airy pattern or aberration-free point spread function for a circular pupil; (a) linear
scale of intensity, (b) logarithmic scale of intensity

then the point spread function for a circular pupil is the Airy pattern, shown in
Fig. 13.1, of which the light intensity distribution is

$$I(\eta) = \left(\frac{2J_1(v)}{v}\right)^2 \Bigg\}$$
$$v = \frac{2\pi na}{\lambda_v R}\cdot\eta \qquad \Bigg\} ; \qquad (13.1)$$

in this equation η is the radial distance from the centre of the pattern, a is the
radius of the pupil, R is the radius of the reference sphere, n is the refractive
index, and λ_v is the vacuum wavelength of the light. There are no physical
units in this equation: it has been normalized so that the intensity of the centre

of the pattern is unity, as is customary in work of this kind. If now an aberrated system is considered, the point spread function changes and becomes very complicated in form. However, if the aberrations are gradually introduced it is found that the initial effects on the point spread function are much the same for any combination of aberration types; these effects are, that the central intensity decreases, the halfwidth of the central maximum does not change and more light appears in the outer rings. Thus we could set as a tolerance level that amount of aberration which produces a just perceptible change of this kind. The first systematic discussion on these lines was given by K. Strehl.[†] A system conforming to the resulting tolerances is said to be diffraction limited, corrected to the Strehl limit or Rayleigh limited corrected.[‡]

Note

Strehl considered the drop in intensity at the centre of the spread function as a fraction of the central intensity in the unaberrated spread function and he suggested a useful tolerance level for this ratio would be 0·8. He called the ratio the *Definitionshelligkeit*, usually rendered as Strehl ratio or Strehl intensity ratio in English texts, and the value of 0·8 for the tolerance level is generally accepted.

Let $W(x, y)$ be the wavefront aberration as a function of coordinates (x, y) in the pupil; this is the function we have already defined in Chapter 7. Then it can be shown[§] that the Strehl ratio is given by

$$1 - \frac{4\pi^2}{\lambda^2} \left[\frac{\iint \{W(x, y)\}^2 \, dx \, dy}{A} - \left(\frac{\iint W(x, y) \, dx \, dy}{A} \right)^2 \right], \qquad (13.2)$$

where A is the area of the pupil and the integrations are taken over the pupil. The term in square brackets can be recognized as the variance of the wavefront aberration, so that the Strehl tolerance level is reached when the variance is $\lambda^2/197$, or in round numbers when the r.m.s. wavefront aberration is $\lambda/14$. The integrals in eqn (13.2) are easily evaluated for the Seidel aberrations with circular pupil; it is best to express the aberrations in polar coordinates (ρ, ϕ) in the pupil as in Section 7.7, but with the pupil radius scaled so that $\rho = 1$ gives the edge of the pupil; the values of the coefficients then give numerically the wavefront aberration at the edge of the pupil. For example, if we consider defocus as an aberration this can be written as $\alpha_1 \rho^2$ [‖] and we find for the Strehl tolerance limit

$$\alpha_1 = \pm 0·25\lambda. \qquad (13.3)$$

---

† *Z. Instrumkde* **22**, 213 (1903).
‡ Lord Rayleigh, in 1878, obtained a tolerance of this kind for defocus considered as an aberration.
§ M. Born and E. Wolf (1964), "Principles of Optics", 4th edition, Pergamon, Oxford.
‖ We shall use Greek letters for the coefficients to remind us that they are for unit radius pupil; thus instead of $a_1, a_2, b_1, \ldots, b_5, \ldots$ etc. as in eqn (7.37) we use $\alpha_1, \alpha_2, \beta_1, \ldots, \beta_5, \ldots$ etc.

This was the case considered by Rayleigh and from it he extrapolated the famous Rayleigh quarter-wavelength rule, that the optical system is substantially perfect if the wavefront can be included between two concentric spherical surfaces $\lambda/4$ apart.

For primary spherical aberration, $\beta_1\rho^4$, the tolerance limit is

$$\beta_1 = \pm 0\cdot 24\lambda. \tag{13.4}$$

However, we noted in Section 7.7 that a better image is obtained if the focus is shifted from the paraxial focal plane and this suggests we should take $W(\rho, \phi)$ as $\alpha_1\rho^2 + \beta_1\rho^4$; then α_1 is chosen to minimize the variance of W and we find for the tolerance level

$$\beta_1 = \pm 0\cdot 95\lambda \quad \text{if } \alpha_1 = -\beta_1. \tag{13.5}$$

It is frequently the case that the optimum focal plane can be chosen and then the tolerance for primary spherical aberration is almost quadrupled.

In the same way we could take a mixture of primary and secondary spherical aberration and find a tolerance on secondary spherical aberration, with defocus and primary spherical aberration as free parameters. For coma we have a similar situation in that the position of maximum light intensity in the comatic point spread function is displaced *laterally* from the Gaussian image point; on the other hand, there is no gain from defocus. The results of several calculations of this kind are collected in Table 13.I, which gives the combination of aberrations and the aberration polynomial at the Strehl tolerance limit.

Table 13.I shows that the tolerances are symmetrical, i.e. the tolerance

TABLE 13.I
Strehl tolerances

Aberration	Aberration polynomial	Coefficients at tolerance limit
Defocus	$\alpha_1\rho^2$	$\alpha_1 = \pm 0\cdot 25\lambda$
Primary spherical aberration (with choice of best focus)	$\alpha_1\rho^2 + \beta_1\rho^4$	$\beta_1 = \pm 0\cdot 95\lambda$ $\alpha_1 = \mp 0\cdot 95\lambda$
Secondary spherical aberration (with choice of best primary and defocus)	$\alpha_1\rho^2 + \beta_1\rho^4 + \gamma_1\rho^6$	$\gamma_1 = \pm 3\cdot 47\lambda$ $\beta_1 = \mp 5\cdot 62\lambda$ $\alpha_1 = \pm 2\cdot 24\lambda$
Primary coma (with choice of lateral shift)	$\alpha_2\rho \cos\phi + \beta_2\rho^3 \cos\phi$	$\beta_2 = \pm 0\cdot 6\lambda$ $\alpha_2 = \mp 0\cdot 4\lambda$
Primary astigmatism (with choice of defocus)	$\alpha_1\rho^2 + \beta_3\rho^2 \cos^2\phi$	$\beta_3 = \pm 0\cdot 35\lambda$ $\alpha_1 = \mp 0\cdot 18\lambda$

limit is numerically the same for a change of sign of the aberration. It would also be found that in all these cases, as in many others, the actual shape of the wavefront conforms fairly well to the Rayleigh quarter-wavelength rule as stated above.

In a given actual optical system the aberrations would occur in combinations, particularly off-axis, which depend in detail on the curvatures etc. and to test for compliance with the Strehl limit it would strictly be necessary to apply eqn (13.2) to the actual aberration polynomial, taking account of the (possibly vignetted) pupil shape. In practice this is not often done; the orders of magnitude of the tolerances in Table 13.I are used as a general guide. If a system has its aberrations within the Strehl tolerance over a certain field, this does not necessarily mean that the point images appear to be perfect; many graphs and photographs showing the light distribution in the point spread function at the Strehl limit for various aberrations have been published[†] and from these it can be seen that the effect of the aberrations is easily visible. Thus occasions could arise when correction to better than the Strehl limit is called for, but for many purposes it is taken to represent substantially perfect correction. The Strehl tolerances can easily be put in terms of ray aberrations and these are sometimes very useful. For example, it is found from Chapter 7 that the defocus tolerance corresponds to a longitudinal focal shift of $\pm 0.5\lambda/u^2$, where u is the convergence angle of the pencil; in other words the *focal range* over which an aberration-free pencil appears to be sharply in focus is λ/u^2. This can be applied to get a tolerance on field curvature and also on longitudinal colour; in the latter case, if all wavelengths are assumed to have equal effects on the detector, the focal range gives the total permissible spread. If an objective has a fixed aperture of diameter $2a$ and if its power is K then $u = Ka$; thus the focal range is proportional to the square of the focal length, whereas the longitudinal colour varies linearly as the focal length (Chapter 10). Therefore if a lens has given diameter and dispersion we can always choose a large enough focal length to ensure that the longitudinal colour spread is within the focal range. For a singlet lens (i.e. not achromatic) of 20 mm diameter the required focal length is easily seen to be of order 1 m. It is for this reason that early seventeenth-century astronomical refractors were made with very long focal ratios, sometimes as much as $f/150$.

The Strehl limit of primary spherical aberration with optimum defocus corresponds to a choice of focal setting at a point halfway from the paraxial focus to the focus of the rays from the edge of the pupil (marginal focus); this contrasts with the purely geometrical optics result that the disc of least confusion is three-quarters of the way from paraxial to marginal focus (Section 7.7). There is no real contradiction here, since the criteria are different, one

† e.g. M. Françon, J. C. Thrierr and M. Cagnet (1962), "Atlas of Optical Phenomena", Springer-Verlag, Berlin.

being based on maximum intensity and the other on the smallest image; also the two criteria are based on different physical theories. Here we make the point that the Strehl intensity calculation for any aberration or combination of aberrations is only valid for small aberrations, i.e. aberrations not much greater than $\lambda/14$ r.m.s.; this is partly because the mathematical approximation involved in obtaining eqn (13.2) becomes invalid and partly because for larger aberrations the form of the point spread function becomes so complicated that the change in intensity at a single point has no overall significance.

The expression given for the Strehl intensity (eqn 13.2) is completely general but simplifications are possible for specific shapes of pupil; if the aberration function $W(x, y)$ is written not as a power series, as we have normally assumed, but as a series of polynomials orthogonal over the pupil, it becomes easier to pick out combinations of a given aberration with defocus, lateral shift and lower order aberrations to give the broadest tolerance, as, for example, for secondary spherical aberration in Table 13.I; in fact, the orthogonal polynomials themselves must represent such combinations. This was done by F. Zernike for the circular pupil and the results are reproduced by Born and Wolf (*loc. cit.*). Again, these apply only to small aberrations and for a pupil of different shape we have to go back to the original formulation of eqn (13.2).

We noted at the beginning of this section the requirements of small convergence angle and large pupil (compared to the wavelength) for the validity of the theory. The case when the convergence angle is small and the pupil is also small nowadays occurs frequently in optical systems used to manipulate and focus laser beams. The parameter which matters is the Fresnel number, $a^2/(\lambda z)$, where a is the pupil radius (or the corresponding laser Gaussian beam parameter w) and z is the focusing distance. Then it is known that for small Fresnel number the centre of curvature of the phasefront in the pupil and the point of maximum intensity in the beam may be well separated axially (see, e.g. H. Kogelnik and T. Li, *Applied Optics*, 5, 1550–1567 (1966)). The corresponding fact for hard-edged pupils, i.e. uniform amplitude over the pupil, has been discussed more recently (see, e.g. Y. Li, *J. Opt. Soc. America*, 72, 770–774 (1982)). In both cases care is necessary in using ray optics to predict the position of maximum intensity along the principal ray for Fresnel numbers less than, say, 10.

13.4   Resolving power and resolution limits

The concept of *resolving power* is frequently used in discussing the quality of images formed by an optical system. In its simplest form the concept is concerned with the capability of an optical system to produce an image of two distinct bright object points which clearly indicates that the object points are separate. Thus for an aberration-free system with a circular pupil if the two points were of equal intensities and were monochromatic the image of each would be an Airy pattern, as in Fig. 13.1 and eqn (13.1), and we should have to consider how closely these images could approach and still be distinguishable as separate point images. To simplify the discussion, it is usual to assume that the object points are *incoherently* illuminated, i.e. that there can be no interference effects between light from the two points, so that the intensity distribution in the combined image is the sum of two Airy patterns at a certain spacing.† Figure 13.2 shows the intensity across the line of centres for some different spacings; it can be seen that the central dip in intensity disappears for spacings less than about 3 units between the centres of the two

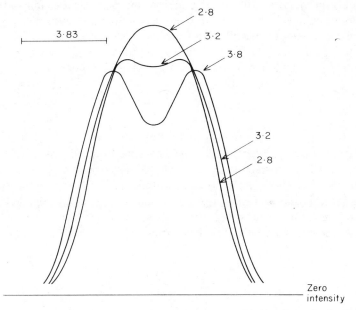

FIG. 13.2. Intensity distribution across two equal Airy patterns added incoherently; the distances between the centres of the pairs of patterns are 2·8, 3·2 and 3·8 diffraction units

† This assumption applies, for example, to two close stars as imaged by a telescope, but it is not necessarily true for, say, pinholes in a metal film seen through a microscope; in this case the illumination is usually partially coherent and for laser illumination there is almost complete coherence. We shall consider only the incoherent case in this book. Likewise, we do not discuss the precise meaning of "monochromatic illumination". These topics, which are closely connected, are discussed by M. Born and E. Wolf (1964), "Principles of Optics", 4th edition, Pergamon, Oxford.

patterns, so that a separation of approximately this value represents in some sense the smallest distinguishable detail. For the moment we call this value k in the units v defined in eqn (13.1); thus from this equation the least distinguishable separation between the geometrical image points is given by

$$\eta_{\min} = \frac{k\lambda_v}{2\pi n(a/R)}. \tag{13.6}$$

There are no firm grounds on which a precise choice of k could be made since even for very small separations the intensity distribution from two points must be slightly different from that for a single point; thus if sufficiently precise measurements could be made of the intensity distribution it would in principle be possible to determine whether two points were being imaged. Actually this approach is impracticable in all but a few special cases, since the demands on the measuring system increase out of all proportion to the gain in resolution and since various sources of random noise mask the very small systematic effects. In practice the value of k generally taken is 3·83, corresponding to the radius of the first dark ring in the Airy pattern, i.e. the first zero of the Bessel function $J_1(v)$; then eqn (13.6) becomes

$$\eta_{\min} = \frac{0·61\lambda_v}{nu}, \tag{13.7}$$

where we have put $u = a/R$. It can be seen by comparing the definition of v in eqn (13.1) with the formula for the Lagrange invariant (eqn (3.20)) that a given value of v represents geometrically conjugate distances in object and image spaces, so that eqn (13.7) can be transferred back to the object space by using the appropriate values of n and u. Also from the optical sine theorem (eqn 6.31) we can write $\sin U$ for u in a space where there is a large convergence angle, giving the familiar formula for "resolving power" in the microscope.†

$$\eta_{\min} = \frac{0·61\lambda_v}{n \sin U}. \tag{13.8}$$

The resolving power or resolution limit so obtained will only apply if the system is substantially perfectly corrected. If appreciable aberrations were present, it would be necessary to use the aberrated point spread function in Fig. 13.2.

Many other resolving power criteria in terms of arrays of bars, etc., have been used, but mainly for experimental testing rather than design calculations.

---

† But, as noted earlier, microscopes are usually used in conditions where the illumination is not completely incoherent.

13.5 Tolerances for non-diffraction-limited systems; definition of the optical transfer function

Many optical systems, e.g. most photographic objectives, do not have anything approaching diffraction-limited aberration correction. The reason may be explained by reference to photographic objectives as an example. A fast objective, say an $f/1\cdot2$ 50 mm lens for a 35 mm camera, could have a theoretical resolution limit of less than 1 μm on the photographic emulsion, according to eqn (13.8); but no ordinary photographic emulsion such as would be used in a 35 mm camera can resolve detail as fine as this and probably 10–20 μm would represent the practical limit. The point of the large aperture is therefore to gain speed, i.e. to be able to use short exposures, not to get high resolution. Such a system is limited in resolution by the detector and there is clearly no point in improving the aberration correction beyond what the detector will respond to. Another example of a detector limited system is an objective for a television camera; the television pick-up tube has between 400 and 1000 lines in its scanning frame, depending on the particular system, and the bandwidth of the channel would normally be such that there is approximately the same resolution along the scan lines as across them; it is obvious then that no detail reproduced by the lens finer than the scan line spacing will be transmitted.

For such detector limited systems it may be convenient simply to use geometrical optics in the form of a spot diagram or some numerical abstract of it, e.g. the r.m.s. distance of the ray intersections from the principal ray, from the Gaussian image point or from some other chosen datum. Many such systems are used in computer-aided optimization of optical design. They are adequate if the aberrations are quite large, so that geometrical optics gives a reasonable picture of the point image. However, there is an intermediate region where the aberrations are above the diffraction-limited level, so that the Strehl ratio system of tolerances cannot be used, but are not so large that purely geometrical optics criteria are valid. The system of tolerances based on the *optical transfer function* (OTF) is useful in this region.

The OTF concept originally developed by analogy with the electrical communications concept of a transfer function for a network or an amplifier; for example, an amplifier will amplify purely sinusoidal signals of different frequencies by different gain factors; the amplification as a function of frequency is the transfer function or response function. In optical systems we introduce a notional sinusoidal test object, which can be thought of as a grating with a sinusoidal transmission profile across the rulings, as in Fig. 13.3. If this is illuminated incoherently it can be shown that the image formed of it by an optical system with *any* aberrations will be of sinusoidal form, as in Fig. 13.4, but with reduced contrast, i.e. the dark parts of the image will

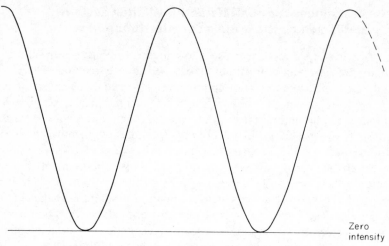

Zero
intensity

FIG. 13.3. A test-chart or "grating" with sinusoidal intensity variation

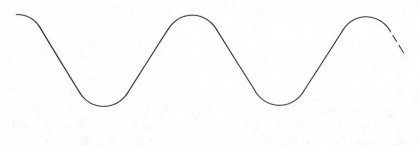

Zero intensity

FIG. 13.4. The image of a sinusoidal grating, showing reduction in contrast

not have zero intensity. The reduction in contrast will depend on the aberrations of the optical system and on the spacing of the grating, i.e. on the spatial frequency; this reduction in contrast as a function of spatial frequency is given by the optical transfer function. The idea was first proposed and developed in detail by Duffieux in 1946† although a few earlier authors, e.g. R. K. Luneburg and W. D. Wright, had hinted at some of the concepts involved. After Duffieux's book several authors took up the theme and image assessment on the basis of the OTF, both by computation from designs and by measurement of actual lens systems, rapidly became an accepted part of

† P. M. Duffieux (1946). "L'intégrale de Fourier et ses applications à l'optique" (published by the author). A second edition, published by Masson et Cie., Paris, appeared in 1970; this is completely rewritten.

applied optics. A review of methods of measurement was given by K. Murata (*Progress in Optics*, V, 199–245 (1966)).

To formulate the OTF concept properly we take axes $O\xi\eta$ in the image plane of an optical system and similarly oriented axes $O\xi_0\eta_0$ in the object plane. However, any point Q with coordinates (ξ_0, η_0) in the object plane is given also a pair of coordinate numbers $(m\xi_0, m\eta_0)$, where m is the transverse magnification of the optical system, and all points in the object plane are denoted by coordinate numbers in this system. This means that an object point and its geometrical image point are denoted by the same "coordinates", (ξ, η), say, which are the *actual* coordinates of the image point. This convention, which we shall use throughout the rest of the chapter without further comment, has advantages of clarity and simplification.

Now let the object have an intensity distribution (as before incoherently illuminated) of the form

$$O(\xi) = a(1 + \cos 2\pi s\xi) \tag{13.9}$$

where a is a constant; in this expression s, which must have the dimensions of a reciprocal length, is the spatial frequency of the sinusoidal grating. Clearly if ξ is measured in, say, millimetres, then s is the number of lines per millimetre† in the grating. Then the result stated above in connection with Figs. 13.3 and 13.4 is that the image of the grating will be of the form

$$O'(\xi) = a'(1 + b \cos (2\pi s\xi + \varepsilon)). \tag{13.10}$$

Here b is a constant less than unity which gives the reduction in contrast in the image and ε is a phase shift which would be expected whenever the point spread function is asymmetric, e.g. because of coma in the optical system. The image takes the form in eqn (13.10) whatever the aberrations of the optical system. The reduction in contrast b and the phase shift ε are functions of the spatial frequency s of the test object; they are expressed together by the OTF, which is defined below. The function $b(s)$ alone gives the drop in contrast or modulation in the image and it is therefore called the modulation transfer function, abbreviated to MTF. For systems in which the spread function is symmetrical the MTF contains all the information about the image quality but in general the phase shift ε is an important factor in the imagery.

The OTF as defined formally is a complex function $\mathscr{L}(s)$ of spatial frequency. It has the property that if the object is as in eqn (13.9) the image will be of the form

$$O'(\xi) = a'(1 + \mathscr{R}(\mathscr{L}(s) e^{2\pi i s\xi})) \tag{13.11}$$

where $\mathscr{R}(\ )$ denotes the real part of the complex quantity in the brackets.

† "Line-pairs" in television technology.

From this it follows by comparison with eqn (13.10) that the MTF is the modulus of the OTF, i.e. $|\mathscr{L}(s)|$ and the phase shift ε is arg $\mathscr{L}(s)$.

So far we have only mentioned that the OTF is a function of spatial frequency but it can easily be seen that this function must depend on the chosen focal plane and, since for all practical optical systems the aberrations vary over the field, the OTF also depends on the position in the field, or the field angle. Furthermore, if the point spread function is not radially symmetrical, as is usually the case except on the axis of a system, the OTF will vary with the azimuth of the lines in the sinusoidal grating involved in defining it. Thus the OTF of a given system is not a function only of spatial frequency but it depends also on the variables mentioned above; alternatively we can say that a given optical system has a large number of different OTFs, corresponding to choice of focal plane, field angle, etc.

Although we have made the obvious point that the aberrations vary with the position in the field, it is on the other hand essential to assume in defining the OTF and computing it that the aberrations are constant over a reasonable number of cycles in the grating; this is the assumption of isoplanatism which is implicit in all theories of imagery of extended objects.

13.6 Formulae for the optical transfer function

In order to be able to calculate the OTF we first need formulae for the point spread function with aberrations. We recall from Section 13.3 that $W(x, y)$ is the wavefront aberration in the pupil. The complex amplitude in the exit pupil produced by a point object is then proportional to exp $\{(2\pi i/\lambda)W(x, y)\}$;[†] we define a *pupil function*, $F(x, y)$, which is equal to exp $\{(2\pi i/\lambda)W(x, y)\}$ over the area of the pupil and is zero elsewhere. Let the distance of the pupil from the image plane be R, i.e. this is the radius of the reference sphere. Then the complex amplitude in the point image at (ξ, η) is, apart from a constant factor,

$$A(\xi, \eta) = \iint\limits_{-\infty}^{\infty} F(x, y)\, e^{-2\pi i(\xi x + \eta y)/\lambda R}\, dx\, dy \qquad (13.12)$$

In this expression the integration is actually over the area of the pupil, but it is customary to define the pupil function $F(x, y)$ formally as above in order to be able to put in infinite limits for the integration. This device is adopted so that the formalism of Fourier transform theory can be used,[‡] e.g. in calculating images of periodic non-sinusoidal objects. It can be seen that if we

---

† Quantities such as complex amplitude are formally defined in books on physical optics, e.g. M. Born and E. Wolf (*loc. cit.*), which also gives proofs of the results we give here.
‡ An excellent heuristic treatment of Fourier transforms is given by R. Bracewell (1965). "The Fourier transform and its applications", McGraw-Hill, New York.

regard $\xi/\lambda R$ and $\eta/\lambda R$ as single variables eqn (13.12) does express $A(\xi, \eta)$ as the two-dimensional Fourier transform of $F(\xi, \eta)$ except that the variables ξ and η are to be re-scaled. This is true for any shape of pupil and any aberrations; varying transmission over the pupil can be incorporated in $F(\xi, \eta)$ and a shift of focus can be produced by adding a focal shift term to the aberration function $W(x, y)$.

The intensity point spread function, which is what is actually measured or seen, is the squared modulus of the amplitude spread function:

$$I(\xi, \eta) = |A(\xi, \eta)|^2. \tag{13.13}$$

In eqns (13.12–13) we have omitted constant factors which would give appropriate dimensions to the physical quantities involved, since these will disappear in the normalizing process we shall use. The appropriate factors can be found in, for example, M. Born and E. Wolf (*loc. cit.*).

We next define the two-dimensional optical transfer function as the normalized inverse Fourier transform of the intensity point spread function

$$\mathscr{L}(s, t) = \frac{\displaystyle\iint_{-\infty}^{\infty} I(\xi, \eta)\, e^{+2\pi i(s\xi + t\eta)}\, d\xi\, d\eta}{\displaystyle\iint_{-\infty}^{\infty} I(\xi, \eta)\, d\xi\, d\eta}. \tag{13.14}$$

Here s and t are spatial frequency components in the ξ and η directions; they have the dimensions of reciprocal length, i.e. "lines" per unit length. The function $\mathscr{L}(s)$ defined in Section 13.4 by eqn (13.11) is, in fact, this same function with the second variable put equal to zero, i.e.

$$\mathscr{L}(s) \equiv \mathscr{L}(s, 0). \tag{13.15}$$

The full two-dimensional function can be used to get the response of a system to grating lines in any azimuth. If we are only interested in spatial frequencies running in, say, the ξ direction, we can put $t = 0$ in eqn (13.14) and obtain

$$\mathscr{L}(s, 0) = \frac{\displaystyle\iint_{-\infty}^{\infty} I(\xi, \eta)\, e^{+2\pi i s\xi}\, d\xi\, d\eta}{\displaystyle\iint_{-\infty}^{\infty} I(\xi, \eta)\, d\xi\, d\eta} \left.\begin{array}{c} \\ \\ \\ \end{array}\right\}$$

$$= \frac{\displaystyle\int_{-\infty}^{\infty} L(\xi)\, e^{+2\pi i s\xi}\, d\xi}{\displaystyle\iint_{-\infty}^{\infty} I(\xi, \eta)\, d\xi\, d\eta} \tag{13.16}$$

where

$$L(\xi) = \int_{-\infty}^{\infty} I(\xi, \eta)\, d\eta. \tag{13.17}$$

The function $L(\xi)$ is the line spread function, i.e. the image of an infinitely narrow incoherently illuminated line. Thus the one-dimensional OTF is the Fourier transform of the line spread function; this result is the principle of frequently used methods of measurement of the OTF. The denominator $\iint I(\xi, \eta)\, d\xi\, d\eta$ in eqns (13.14) and (13.16) is the normalizing factor referred to above. It ensures that $\mathscr{L}(0, 0) \equiv 1$, which is obviously necessary, since there should be no loss in contrast in the image of an object of uniform intensity.

An alternative expression for the OTF can be obtained in terms of the auto-correlation of the pupil function:

$$\mathscr{L}(s, t) = \frac{\displaystyle\iint_{-\infty}^{\infty} F(\lambda Rs + x, \lambda Rt + y)F^*(x, y)\, dx\, dy}{\displaystyle\iint_{-\infty}^{\infty} |F(x, y)|^2\, dx\, dy}. \tag{13.18}$$

Again, the denominator is a normalizing factor to make $\mathscr{L}(0, 0) \equiv 1$. The range of integration in the autocorrelation is only formally infinite; if the pupil is drawn as in Fig. 13.5 it can be seen that the integral is only non-zero over the area common to the pupil and its twin with origin shifted to $(-\lambda Rs,$

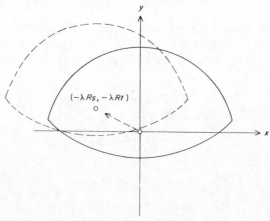

FIG. 13.5. The sheared pupils used to calculate the OTF as an autocorrelation; the spatial frequency components are (s, t) if the components of shear are $(\lambda Rs, \lambda Rt)$

$-\lambda Rt$). This expression shows immediately that $\mathscr{L}(s, t)$ must vanish identically for spatial frequencies exceeding certain limits; these are given by the values of λRs or λRt needed to shear the twin pupil completely clear of the first one. Thus for a circular pupil of radius a this cut-off spatial frequency is

$$s_{\max}, t_{\max} = \frac{2a}{\lambda R}. \qquad (13.19)$$

Alternatively, this can be put in the form that the limiting distinguishable separation between the bars or lines in the sinusoidal grating is

$$\frac{0\cdot 5\lambda}{(a/R)} \qquad (13.20)$$

which bears obvious similarities to eqn (13.6).

Both the approaches described above have been used for computing the OTF of designed systems. An essential preliminary is to have adequate ray tracing results to give the pupil function $F(x, y)$ in enough detail, often with $W(x, y)$ as an aberration polynomial. Then in the first method the complex amplitude point spread function is calculated (eqn (13.2)), it is squared to give the intensity point spread function (eqn (13.13)) and Fourier transformed to give the OTF (eqn (13.14)). In the second method the autocorrelation of the pupil function is taken immediately (eqn (13.18)). The latter would seem to be more direct, but in fact there is still at the time of writing no definite agreement on the best technique. It clearly depends on the computing facilities available, both machines and programs.

From eqn (13.18) and the construction in Fig. 13.5 it can be shown that the OTF for an aberration-free circular pupil of radius a is given by

$$\mathscr{L}(s) = \frac{2}{\pi} \arccos \frac{x}{2a} - \frac{x}{\pi a} \sqrt{1 - \frac{x^2}{4a^2}} \left. \begin{array}{c} \\ \\ \\ \\ \end{array} \right\}$$
$$= \frac{2}{\pi} \arccos \frac{\lambda Rs}{2a} - \frac{\lambda Rs}{\pi a} \sqrt{1 - \frac{\lambda^2 R^2 s^2}{4a^2}} \qquad (13.21)$$

This is written as a function of one variable, the spatial frequency s, since it is obviously the same function for gratings in all azimuths. It is plotted in Fig. 13.6.

So far we have made the implicit assumptions that the convergence angle and the field angle are both small. The first assumption is essential to scalar diffraction theory; however, it is quite practicable to define the OTF at large field angles. We have to define pupil coordinates in the pupil plane transverse to the optical axis, as in Fig. 13.7, and image plane coordinates in a parallel plane. The pupil function $F(x, y)$ is defined in terms of these coordinates,

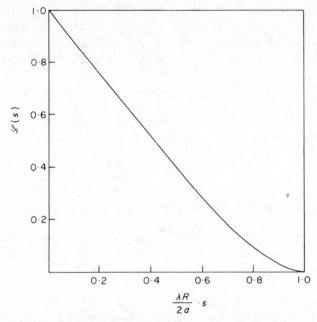

FIG. 13.6. The OTF for an aberration-free circular pupil; the transfer function is plotted against the normalized spatial frequency, $s/(2a/\lambda R)$

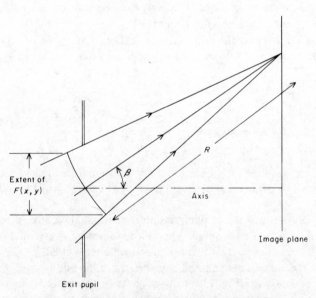

FIG. 13.7. Calculating the OTF for a pencil at a large field angle

paying due regard to vignetting and pupil aberrations. It is then found that the complex amplitude point spread function is

$$A(\xi, \eta \cos \beta) = \int\limits_{-\infty}^{\infty}\int F(x, y)e^{-2\pi i(\xi x + \eta y)/\lambda R} \, dx \, dy \qquad (13.22)$$

where β is the field angle. The two-dimensional transfer function is

$$\mathcal{L}(s, t) = \dfrac{\int\limits_{-\infty}^{\infty}\int |A(\xi, \eta \cos \beta)|^2 \, e^{2\pi i(s\xi + t\eta)} \, d\xi \, d\eta}{\int\limits_{-\infty}^{\infty}\int |A|^2 \, d\xi \, d\eta} \qquad (13.23)$$

but when it is put in terms of the autocorrelation over the pupil the cosine term disappears:

$$\mathcal{L}(s, t) = \dfrac{\int\limits_{-\infty}^{\infty}\int F(\lambda Rs + x, \lambda Rt + y)F^*(x, y) \, dx \, dy}{\int\int |F(x, y)|^2 \, dx \, dy}. \qquad (13.24)$$

In each case the spatial frequencies are measured in the image plane transverse to the optical axis, *not* transverse to the principal ray. Also, as explained above, the pupil coordinates are also in a plane transverse to the axis; thus as can be seen from Fig. 13.7 the pupil function $F(x, y)$ is non-zero over the domain indicated, although the rays bounding the pencil cut the pupil plane in a different outline.

13.7 The OTF in the geometrical optics approximation

For some purposes it may be thought adequate to use the approximation of geometrical optics to calculate an "OTF"; this is done by representing the light intensity in the image by the density of ray intersections in the image plane. From eqn (7.47) the ray intersection density can be written

$$I_g(\xi, \eta) = C \frac{\partial(x, y)}{\partial(\xi, \eta)}, \qquad (13.25)$$

where (x, y) are pupil coordinates, (ξ, η) are image plane coordinates and C is a proportionality constant. We have called this $I_g(\xi, \eta)$, a function of the image plane coordinates, since it can be taken to represent the point spread function according to geometrical optics. Equation (13.25) is valid at finite field angles provided the two coordinate planes are perpendicular to the optical axis, as explained at the end of the last section.

We can now write the geometrical optics OTF as the Fourier transform of the point spread function, just as in eqn (13.16),

$$\mathcal{L}_g(s, t) = K \int\!\!\!\int\limits_{-\infty}^{\infty} \frac{\partial(x, y)}{\partial(\xi, \eta)} \, e^{2\pi i(\xi s + \eta t)} \, d\xi \, d\eta, \qquad (13.26)$$

where K is a normalizing constant; if we use the Jacobian to change the variables of integration and then write ξ and η in terms of the aberration function we obtain

$$\mathcal{L}_g(s, t) = K \int\!\!\!\int\limits_{-\infty}^{\infty} \exp\left[2\frac{\pi i R}{n}\left(s\frac{\partial W}{\partial x} + t\frac{\partial W}{\partial y} \right) \right] dx \, dy. \qquad (13.27)$$

It is easily seen that the normalizing constant is the reciprocal of the projected area of the wavefront, as in Fig. 13.7. These results are quoted by K. Miyamoto† in a review which discusses several criteria of image quality; Miyamoto gives examples of comparisons between OTF calculated according to geometrical optics and according to diffraction theory. Obviously the "OTF" calculated in this way can give very wrong results for small aberrations, e.g. it can give non-zero values for spatial frequencies beyond the cut-off limit given by eqn (13.19). Thus sometimes one makes use of the geometrical optics OTF by taking its product with the diffraction OTF for zero aberration, i.e. that given by Fig. 13.6; this curious approach seems in practice to give acceptable results.

13.8 Incoherently illuminated lines and edges as test objects

Occasionally a narrow bright line has been used as the basis of design tolerances; this would be appropriate in, for example, the design of spectroscopic systems. We have already noted the expression for the line spread function (eqn (13.17)); some parameter extracted from the shape, e.g. the width at half maximum intensity or the central intensity, is used to set a tolerance on the aberrations. A system analogous to the Strehl system for the point spread function was given in 1963.‡ The light distribution in the image of an opaque edge along the η-axis with its dark side given by negative values of ξ can be

---

† "Progress in Optics", Vol. I (1961), pp. 33–67 (E. Wolf, ed.) North-Holland, Amsterdam.
‡ W. T. Welford (1963). *Optica Acta* **10**, 121–127.

shown to be given by

$$E(\xi) = \int_{-\infty}^{\xi} L(\xi') \, d\xi', \qquad (13.28)$$

where $L(\xi)$ is the image of a bright line along the η-axis. An edge image has been used as a quality criterion for photographic emulsions, taking either the maximum gradient in the image or the distance between, say, the 10% and 90% intensity points. The gradient of intensity in the incoherent edge image always has the same sign, so that there are no maxima or minima in the image, whatever the aberrations or pupil shape. Thus it shares the pleasing property of the sinusoidal grating, that the fine diffraction detail in the point spread function is smoothed out.

The line, the edge and the OTF are closely related; as seen in Section 13.6, the line image is the Fourier transform of the OTF; also from eqn (13.28) the intensity in the line image is the gradient of the intensity in the edge image, suitably scaled; finally it can be shown that the area of the graph of $\mathscr{R}(\mathscr{L}(s, 0))$, the real part of the OTF, is proportional to the intensity of the line spread function at the origin of ξ and therefore also to the gradient at the centre of the edge image. Thus the edge gradient is an average of the OTF.

13.9 Optical tolerances and image assessment

In the earlier sections of this chapter we have indicated where the different tolerance systems apply when this is fairly clearcut. Unfortunately in many cases the best choice of optical tolerances is by no means clear; in this section we discuss this more general problem, mainly to point out the difficulties rather than give the answers.

The difficulties generally arise in connection with non-diffraction-limited systems. If there is some clear basis for applying an OTF based tolerance, as in a television camera lens, then matters are relatively simple, but even here the question of the relative importance to be attached to a phase shift in the OTF has not been examined in the literature. The tolerance specifications usually seem to be written in terms of MTF only and it is therefore probable that the optical designers have ignored phase shifts, on the very reasonable grounds that they have enough to worry about already. If there is no clear basis for an OTF tolerance but yet the system is clearly detector limited we can try to bring in the detector response, but in fact most image-recording detectors—photographic emulsion, TV channel, human eye, dry copying process, etc.—are very nonlinear in response, so that they cannot strictly be assigned an OTF: anyway, a given system may be used with several different detectors.

A more fundamental difficulty is that for many optical systems we do not know what kind of images are to be formed and this is a severe limitation in assigning optical tolerances. In certain cases the kind of image is known and this information can be used in tolerancing; for instance spectrograph objectives and cathode-ray tube relay lenses form line images, flying spot relay lenses form point images, reflecting astronomical telescopes usually form point images (refractors are usually used for low-contrast extended objects such as planets). In these cases it is generally possible to put a sensible tolerance on the line or point spread function as a basis for optical design.

In the vast majority of cases the choice of tolerance system is made by the optical designer on the basis of previous experience and since in so many cases the ultimate test on the optical system will be the users' approval based on subjective judgements this is probably the best that can be done.

When it comes to testing a manufactured system there are conflicting requirements for the tolerance system. The tests which involve assessment of an image, OTF, line spread function, etc. are attractive to the prospective user for obvious reasons; however, it is not possible to calculate aberrations from the results of such tests, so that they do not give a direct indication to the designer of what may have gone wrong in the realization of his design. Also these tests depend, sometimes very critically, on the mode of illumination, on the spectral distribution of the illumination and on choice of focus (it is not possible to calculate the OTF at a second focal setting from a knowledge of it at one focal position). On the other hand a direct measurement of aberrations tells the designer a lot more about how the real system and the design differ, although the aberrations may mean little to the prospective user. Also if the aberrations are known at a single focal position the effects of defocus can be found and they can be used to compute image assessment data like the OTF.

From the above it will be seen that there is no single simple answer to the problem of choosing tolerances; the designer should be flexible in his outlook and should have programs available to cover all contingencies.

Appendix A   Summary of the Main Formulae

In this appendix the principal formulae are collected together with references to the original equations in the book.

A.1 Gaussian optics

Newton's conjugate distance equation

$$zz' = ff'. \tag{1.8}$$

Transverse magnification

$$m = \frac{\eta'}{\eta} = -\frac{f}{z} = -\frac{z'}{f'} = -\frac{l'f}{lf'}. \tag{1.6), (1.7), (1.13}$$

Conjugate distance equations referred to principal points.

$$\frac{f'}{l'} + \frac{f}{l} = 1, \tag{1.11}$$

$$\frac{n'}{l'} - \frac{n}{l} = K. \tag{3.15}$$

Snell's law

$$n' \sin I' = n \sin I, \tag{2.2}$$

$$n'(\mathbf{r}' \wedge \mathbf{n}) = n(\mathbf{r} \wedge \mathbf{n}). \tag{2.3}$$

Ratio of focal lengths

$$\frac{n'}{f'} = -\frac{n}{f} = K. \tag{3.14}$$

Lagrange invariant

$$n'u'\eta' = nu\eta = -nh\beta = H.$$ (3.20), (3.29)

Angular magnification of afocal systems

$$\frac{\beta'}{\beta} = \frac{nh}{n'h'}.$$ (3.31)

Longitudinal magnification of afocal systems

$$\frac{l'}{l} = \frac{n'h'^2}{nh^2}.$$ (3.33)

A single refracting surface

$$\frac{n'}{l'} - \frac{n}{l} = (n' - n)c = K,$$ (3.38)

$$f = \frac{-n}{(n' - n)c}, \quad f' = \frac{n'}{(n' - n)c}.$$ (3.39)

Combination of two systems

$$K = K_1 + K_2 - \frac{d}{n_1'} K_1 K_2,$$ (3.48)

$$\delta = \frac{n}{n_1'} \frac{dK_2}{K}, \quad \delta' = -\frac{n'}{n_1'} \frac{dK_1}{K},$$ (3.52), (3.51)

Thick lens in medium n_0

$$K = (n - n_0)\left\{c_1 - c_2 + \frac{n - n_0}{n} dc_1 c_2\right\}.$$ (3.53)

Thin lens

$$K = (n - 1)(c_1 - c_2).$$ (3.56)

Two thin lenses

$$K = K_1 + K_2 - dK_1 K_2.$$ (3.57)

Two systems referred to the interfocal distance

$$K = \frac{-g}{n_1'} K_1 K_2,$$ (3.39)

$$\Delta = \frac{nK_2}{K_1 K}, \quad \Delta' = -\frac{n'K}{K_2 K}$$ (3.62) (3.61)

Paraxial raytracing

$$n'u' - nu = -hK,$$ (3.66)

$$h_{+1} = h + d'u'.$$ (3.68)

Refraction invariant

$$A = n(hc + u)$$ (6.2)

A.2 Finite raytracing

Transfer between spherical surfaces

$$x_0 = x_{-1} + \frac{L}{N}(d - z_{-1}),$$

$$y_0 = y_{-1} + \frac{M}{N}(d - z_{-1}),$$
(4.4)

$$F = c(x_0^2 + y_0^2),$$ (4.9)

$$G = N - c(Lx_0 + My_0),$$ (4.10)

$$\Delta = \frac{F}{G + \sqrt{G^2 - cF}},$$ (4.12)

$$\left.\begin{array}{l} x = x_0 + L\Delta \\ y = y_0 + M\Delta \\ z = N\Delta \end{array}\right\}.$$ (4.5)

Refraction through a spherical surface

$$\cos I = +\sqrt{G^2 - cF},$$ (4.16)

$$n' \cos I' = +\{n'^2 - n^2(1 - \cos^2 I)\}^{\frac{1}{2}},$$ (4.17)

$$K = c(n' \cos I' - n \cos I),$$ (4.15)

$$\left.\begin{array}{l} n'L' = nL - Kx \\ n'M' = nM - Ky \\ n'N' = nN - Kz + n' \cos I' - n \cos I \end{array}\right\}.$$ (4.14)

Meridian rays by trigonometrical formulae

$$\left.\begin{array}{l} \sin I = -\dfrac{(L - r)}{r} \sin U \\ n' \sin I' = n \sin I \\ I' - U' = I - U \\ L' - r = \dfrac{r \sin I'}{\sin U'} \end{array}\right\}.$$ (4.43)

A.3 Primary monochromatic aberrations

Eccentricity parameter

$$\frac{\bar{h}}{h} = HE, \tag{6.7}$$

$$E_{+1} - E = \frac{-d'}{n'hh_{+1}}. \tag{6.16}$$

Parameters of the principal ray in terms of the eccentricity parameter (Seidel difference formulae)

$$\bar{A} = -\frac{H}{h}(1 - AhE), \tag{6.9}$$

$$n\bar{u} = -\frac{H}{h}(1 - nuhE), \tag{6.11}$$

$$\bar{h} = hHE. \tag{6.7}$$

The Hamilton expansion

$$\begin{aligned}
W(x^2 + y^2, y\eta, \eta^2) = {} & a_1(x^2 + y^2) + a_2 y\eta + a_3\eta^2 + b_1(x^2 + y^2)^2 \\
& + b_2 y\eta(x^2 + y^2) + b_3 y^2\eta^2 + b_4\eta^2(x^2 + y^2) \\
& + b_5 y\eta^3 + b_6\eta^4 + \cdots.
\end{aligned} \tag{7.37}$$

Seidel sums

$$\left.\begin{aligned}
S_I &= -\sum A^2 h\Delta\left(\frac{u}{n}\right) \\[2mm]
S_{II} &= -\sum \bar{A} Ah\Delta\left(\frac{u}{n}\right) \\[2mm]
S_{III} &= -\sum \bar{A}^2 h\Delta\left(\frac{u}{n}\right) \\[2mm]
S_{IV} &= -\sum H^2 c\Delta\left(\frac{1}{n}\right) \\[2mm]
S_V &= -\sum \left\{\frac{\bar{A}^3}{A} h\Delta\left(\frac{u}{n}\right) + \frac{\bar{A}}{A} H^2 c\Delta\left(\frac{1}{n}\right)\right\}
\end{aligned}\right\} . \tag{8.41}$$

Total primary wavefront aberration

$$W(x_p, y_p, \eta) = \tfrac{1}{8}S_I \frac{(x_p^2 + y_p^2)^2}{h_p^4} + \tfrac{1}{2}S_{II} \frac{y_p(x_p^2 + y_p^2)}{h_p^3} \cdot \frac{\eta}{\eta_{max}} + \tfrac{1}{2}S_{III} \frac{y_p^2}{h_p^2} \cdot \frac{\eta^2}{\eta_{max}^2}$$

$$+ \tfrac{1}{4}(S_{III} + S_{IV}) \frac{(x_p^2 + y_p^2)}{h_p^2} \cdot \frac{\eta^2}{\eta_{max}^2} + \tfrac{1}{2}S_V \frac{y_p}{h_p} \cdot \frac{\eta^3}{\eta_{max}^3}. \tag{8.42}$$

Primary transverse ray aberrations

Spherical aberration: $\dfrac{S_{\mathrm{I}}}{2nu}$. $\qquad$ (8.48)

Sagittal coma: $\dfrac{S_{\mathrm{II}}}{2nu}$. $\qquad$ (8.50)

Tangential coma: $\dfrac{3S_{\mathrm{II}}}{2nu}$. $\qquad$ (8.51)

Length of astigmatic focal line: $\left|\dfrac{2S_{\mathrm{III}}}{nu}\right|$. $\qquad$ (8.52)

Distortion: $\delta\eta' = \dfrac{S_{\mathrm{V}}}{2nu}$. $\qquad$ (8.62)

Fractional distortion: $\dfrac{\delta\eta'}{\eta'} = \dfrac{S_{\mathrm{V}}}{2H}$. $\qquad$ (8.63)

Differential distortion: $\dfrac{d\eta'}{d\eta} = m\left(1 + \dfrac{3S_{\mathrm{V}}}{2H}\right)$. $\qquad$ (8.64)

Primary longitudinal ray aberrations

Spherical aberration: $\dfrac{S_{\mathrm{I}}}{2nu^2}$. $\qquad$ (8.49)

Distance between t and s foci: $\dfrac{S_{\mathrm{III}}}{nu^2}$. $\qquad$ (8.53)

Gaussian image plane to t-focus:

$$\dfrac{3S_{\mathrm{III}} + S_{\mathrm{IV}}}{2nu^2}.$$ $\qquad$ (8.54)

Gaussian image plane to s-focus:

$$\dfrac{S_{\mathrm{III}} + S_{\mathrm{IV}}}{2nu^2}.$$ $\qquad$ (8.55)

Gaussian image plane to astigmatic disc of least confusion:

$$\dfrac{2S_{\mathrm{III}} + S_{\mathrm{IV}}}{2nu^2}.$$ $\qquad$ (8.56)

Gaussian image plane to Petzval surface:

$$\dfrac{S_{\mathrm{IV}}}{2nu^2}.$$ $\qquad$ (8.57)

Image surface curvatures:

Tangential $\qquad\qquad \dfrac{n(3S_{III} + S_{IV})}{H^2}$.

$\hspace{10cm}$ (8.58)

Sagittal $\qquad\qquad \dfrac{n(S_{III} + S_{IV})}{H^2}$.

$\hspace{10cm}$ (8.59)

Least confusion $\qquad \dfrac{n(2S_{III} + S_{IV})}{H^2}$.

$\hspace{10cm}$ (8.60)

Petzval $\qquad\qquad \dfrac{nS_{IV}}{H^2}$.

$\hspace{10cm}$ (8.61)

Primary angular ray aberrations:

Spherical aberration: $\qquad \dfrac{\beta S_I}{2H}$.

$\hspace{10cm}$ (8.65)

Tangential coma: $\qquad \dfrac{3\beta S_{II}}{2H}$.

$\hspace{10cm}$ (8.66)

Sagittal coma: $\qquad \dfrac{\beta S_{II}}{2H}$.

$\hspace{10cm}$ (8.67)

Distortion $\qquad\qquad \dfrac{\beta S_V}{2H}$.

$\hspace{10cm}$ (8.68)

Stop shift effects

If $\varepsilon = H\,\delta E$, the change in HE, then

$$\delta \overline{A} = \varepsilon A \hspace{6cm} (8.73)$$

$$\left.\begin{aligned}
\delta S_{II} &= \varepsilon S_I \\
\delta S_{III} &= 2\varepsilon S_{II} + \varepsilon^2 S_I \\
\delta S_V &= \varepsilon(3S_{III} + S_{IV}) + 3\varepsilon^2 S_{II} + \varepsilon^3 S_I
\end{aligned}\right\}. \hspace{2cm} (8.74)$$

Thin lens primary aberrations

Shape or bending variable; $\quad B = \dfrac{n_0(n-1)(c_1 + c_2)}{K} = \dfrac{c_1 + c_2}{c_1 - c_2}$. (12.2)

Conjugate variable:

$$C = \dfrac{n_0(u_1 + u'_2)}{hK} = \dfrac{u_1 + u'_2}{u_1 - u'_2}. \hspace{3cm} (12.2)$$

Seidel sums, with remote stop

$$S_I = \frac{h^4 K^3}{4n_0^2}\left\{\left(\frac{n}{n-1}\right)^2 + \frac{n+2}{n(n-1)^2}\left(B + \frac{2(n^2-1)}{n+2}C\right)^2 - \frac{n}{n+2}C^2\right\}$$

$$S_{II} = \frac{-h^2 K^2 H}{2n_0^2}\left\{\frac{n+1}{n(n-1)}B + \frac{(2n+1)}{n}C\right\} + HE.S_I$$

$$S_{III} = \frac{H^2 K}{n_0^2} - HE\cdot\frac{h^2 K^2 H}{n_0^2}\left\{\frac{n+1}{n(n-1)}B + \frac{(2n+1)}{n}C\right\} + (HE)^2 S_I$$

$$S_{IV} = \frac{H^2 K}{n_0^2 n}$$

$$S_V = HE\cdot\frac{H^2 K}{n_0^2}\left\{3 + \frac{1}{n}\right\}$$
$$- 3(HE)^2\frac{h^2 K^2 H}{2n_0^2}\left\{\frac{n+1}{n(n-1)}B + \frac{(2n+1)}{n}C\right\} + (HE)^3 S_I$$

$$C_I = \frac{h^2 K}{V}$$

$$C_{II} = HE.C_I$$

(12.17)

Plane parallel plates

$$S_I = -\frac{n_0(n^2-1)}{n^3}u^4 d$$

$$S_{II} = \frac{\bar{u}}{u}S_I$$

$$S_{III} = \left(\frac{\bar{u}}{u}\right)^2 S_I$$

$$S_V = \left(\frac{\bar{u}}{u}\right)^3 S_I$$

(12.21)

A.4 Total aberrations

Transverse ray aberrations from the wavefront aberrations

$$\delta\xi = -\frac{R}{n}\frac{\partial W}{\partial x}$$
$$\delta\eta = -\frac{R}{n}\frac{\partial W}{\partial y}$$

(7.15)

Wavefront aberration from transverse ray aberrations

$$W_B - W_A = \int_A^B \{\delta\xi \, dx + \delta\eta \, dy\}. \tag{7.16}$$

Linear coma

$$W_{coma}(a' \sin\phi, a' \cos\phi) = H \cos\phi \left\{\frac{(l' - \bar{l}' - \delta l') \sin U'}{(l' - \bar{l}')} \frac{\sin U'}{u'} - \frac{\sin U}{u}\right\}. \tag{9.39}$$

Offence against the sine condition

$$\frac{\sin U}{u} \cdot \frac{u'}{\sin U'} \left(\frac{l' - \bar{l}'}{l' - \bar{l}' - \delta l'}\right) - 1. \tag{9.40}$$

Abbe sine condition

$$\frac{\sin U'}{u'} = \frac{\sin U}{u}. \tag{9.43}$$

Thomas Young's astigmatism formulae (s and t formulae)

$$\frac{n' \cos^2 I'}{t'} - \frac{n \cos^2 I}{t} = c(n' \cos I' - n \cos I), \tag{9.69}$$

$$\frac{n'}{s'} - \frac{n}{s} = c(n' \cos I' - n \cos I). \tag{9.72}$$

A.5 Chromatic aberrations

Dispersion
$$V = \frac{n - 1}{\delta n}. \tag{10.6}$$

Achromatic doublet
$$\frac{K_1}{V_1} + \frac{K_2}{V_2} = 0. \tag{10.3}$$

Conrady "D minus d" formula

$$\delta_\lambda W = \sum \delta n(D - d) \tag{10.4}$$

Primary chromatic aberrations

$$C_1 = \sum Ah\Delta\left(\frac{\delta n}{n}\right), \tag{10.19}$$

$$C_{II} = \sum \bar{A}h\Delta\left(\frac{\delta n}{n}\right), \tag{10.22}$$

$$\delta C_{II} = \varepsilon C_I. \tag{10.23}$$

Chromatic focal shift

$$\delta_\lambda l_k' = \frac{1}{n_k' u_k'^2} \cdot C_I.$$
(10.24)

Chromatic change in image height

$$\delta_\lambda \eta_k' = -\frac{1}{n_k' u_k'} \cdot C_{II}.$$
(10.25)

Chromatic aberration of thin lenses

$$C_I = \frac{h^2 K}{V}$$
(12.10)

$$C_{II} = 0 \qquad \text{(stop at the lens)}$$
(12.11)

$$C_{II} = HE.C_I \qquad \text{(remote stop)}$$
(12.17)

Plane parallel plates

$$\left.\begin{aligned}
C_I &= -\frac{n_0}{n^2} u^2 d.\delta n \\
C_{II} &= -\frac{n_0}{n^2} u\bar{u} d.\delta n
\end{aligned}\right\}.$$
(12.21)

A.6 Holographic optical elements and gratings

Vector raytracing equation for holograms

$$\mathbf{n} \wedge (\mathbf{r}_0' - \mathbf{r}_r') = \frac{m\lambda'}{\lambda} \cdot \mathbf{n} \wedge (\mathbf{r}_0 - \mathbf{r}_r).$$
(5.33)

Resolved parts of eqn (5.33)

$$\left.\begin{aligned}
L_0' - L_r' &= \frac{m\lambda'}{\lambda}(L_0 - L_r) \\
M_0' - M_r' &= \frac{m\lambda'}{\lambda}(M_0 - M_r).
\end{aligned}\right\}$$
(5.34)

The grating equation

$$\sin \alpha + \sin \alpha' = m\lambda/\sigma.$$
(11.9)

Aberration expansion for gratings

$$W = \text{const.} + \tfrac{1}{2}(T + T')y^2 + \tfrac{1}{2}(S + S')x^2 + \frac{1}{2}\frac{m\lambda}{\sigma}\frac{\xi'^2}{r'^2}y + \cdots$$
(11.11)

with

$$S = \frac{1}{r} - \frac{\cos \alpha}{R}, \qquad T = \frac{\cos^2 \alpha}{r} - \frac{\cos \alpha}{R}, \qquad \text{etc.} \qquad (11.10)$$

Primary spherical aberration of a plane holographic lens

$$W = \tfrac{1}{8} y^4 \left\{ \frac{1}{m} \left(\frac{1}{l_0'^3} - \frac{1}{l_r'^3} \right) - \frac{\lambda'}{\lambda} \left(\frac{1}{l_0^3} - \frac{1}{l_r^3} \right) \right\}. \qquad (11.17)$$

Off-axis primary aberrations of a plane holographic lens

$$-\frac{1}{2l_0'} \left(\frac{1}{l_0'^2} - \frac{1}{l_r'^2} \right) \eta_0 \, y(x^2 + y^2) + \frac{3}{4l_0'^2} \left(\frac{1}{l_0'} - \frac{1}{l_r'} \right) \eta_0^2 \, y^2$$

$$+ \frac{1}{4l_0'^2} \left(\frac{1}{l_0'} - \frac{1}{l_r'} \right) \eta_0^2 x^2. \qquad (11.20)$$

Astigmatism along a principal ray for a holographic optical element on a non-plane substrate

$$\frac{1}{m\lambda'} \left\{ c_t(\cos I_0' - \cos I_r') - \frac{\cos^2 I_0'}{t_0'} + \frac{\cos^2 I_r'}{t_r'} \right\}$$

$$- \frac{1}{\lambda} \left\{ c_t(\cos I_0 - \cos I_r) - \frac{\cos^2 I_0}{t_0} + \frac{\cos^2 I_r}{t_r} \right\} = 0. \qquad (11.24)$$

$$\frac{1}{m\lambda'} \left\{ c_s(\cos I_0' - \cos I_r') - \frac{1}{s_0'} + \frac{1}{s_r'} \right\}$$

$$- \frac{1}{\lambda} \left\{ c_s(\cos I_0 - \cos I_r) - \frac{1}{s_0} + \frac{1}{s_r} \right\} = 0. \qquad (11.25)$$

Condition for aplanatism for a holographic lens on a spherical substrate

$$\frac{1}{l'} + \frac{1}{l} = \frac{1}{r}. \qquad (11.29)$$

Appendix B   Symbols

The main symbols used are collected in this appendix with references to the sections where they are defined.

$A(\xi, \eta)$	complex amplitude point spread function	13.6
a_1–a_3, b_1–b_6, etc.	coefficients in the Hamilton expansion	7.6
B	shape or bending variable for thin lenses	12.1
bfl	back focal length	3.9
C	conjugate variable for thin lenses	12.1
C_I, C_II	primary chromatic aberration sums	10.5
c	curvature	3.1
D	length of ray segment	7.5
d	axial thickness or separation	3.8
E	eccentricity parameter	5.3
E	the eikonal	7.2
efl	equivalent focal length	3.2
F, F$'$	principal foci	1.3, 3.2
F, G	auxiliary quantities for raytracing	4.3
$F(x, y)$	pupil function	13.6
f, f'	principal focal lengths	1.3, 3.2
G	aspheric coefficient	8.6
g	axial separation between foci	3.8
H	Lagrange invariant	3.3
$\mathscr{H}$	skew invariant	6.4
h	paraxial incidence height	3.3
I	angle of incidence	2.2

$I(\xi, \eta)$	Intensity point spread function	13.6
K	power	3.2
$\mathcal{K}$	generalized power	4.4
L	finite intersection length	4.9
(L, M, N)	direction cosines of a ray	4.2
$\mathcal{L}(s, t)$	optical transfer function	13.5
l, l'	conjugate distances from principal points	1.3, 3.2
m	transverse magnification	1.2, 1.3
N, N'	nodal points	1.3, 3.2
n	refractive index	2.1
$\boldsymbol{n}$	unit normal vector	2.2
P	relative partial dispersion	10.3
P, P'	principal points	1.3, 3.2
R	radius of reference sphere	7.2
r	radius of curvature	3.9
$\boldsymbol{r}$	unit vector along a ray	2.2
S	reference sphere	7.2
S	shortest distance between skew rays	6.4
$S_{\mathrm{I}}, S_{\mathrm{II}}, \ldots, S_{\mathrm{V}}$ Seidel sums		8.2
S_{VI}	spherical aberration of pupil imagery	8.7
s	sagittal intersection length	9.8
s, t	spatial frequency components	13.6
t	tangential intersection length	9.8
U	finite convergence angle	4.9
u	paraxial convergence angle	3.3
V	Hamilton's point characteristic function	7.2
V	reciprocal dispersive power	10.2
W	optical path length	2.3
W	wavefront aberration	7.2
z, z'	conjugate distances from principal foci	1.3, 3.2
(α, β, γ)	direction cosines of a normal	4.2
β	field angle	3.3
$\Delta(\ )$	increment on refraction	8.2
Δ, Δ'	coordinates of foci	3.8
δ, δ'	coordinates of principal points	3.8
ε	change in HE for a stop shift	8.5
ε	asphericity parameter of a quadric of revolution	4.8
η, η'	object or image height	1.3
θ	angle between skew rays	6.4
ρ, ϕ	polar coordinates in the pupil	7.7
Σ	geometrical wavefront	2.1

Appendix C   Examples

1 The cornea (front surface) of the human eye has a radius of curvature of about 8 mm. Assuming as a crude approximation that the interior of the eye is a uniform medium of refractive index 1·34 calculate (a) the two focal lengths of the eye, (b) its power and (c) the positions of the nodal points.

2 An afocal system is to be made of two thin lenses of focal lengths 100 mm and 10 mm respectively. Find the positions of the image of an object placed at the following distances from the 100 mm lens: -1000 mm, -200 mm, -100 mm and 0 mm. Sketch ray diagrams showing how the images are formed if the aperture stop is at the 100 mm lens.

3 A lens is to have an equivalent focal length of 100 mm and its centre thickness is to be 10 mm. If the refractive index of the material is 1·523 calculate the curvatures (a) for an equiconvex lens and (b) for a planoconvex lens. Find the positions of the principal planes in each case.

4 A lens has the form of a complete sphere of radius r. Show that its power is

$$\frac{2(n-1)}{nr}.$$

For what refractive index will an object at infinity be imaged at the rear surface of this lens?

5 The reflection factor for light polarized with the *E*-vector parallel to the plane of incidence is

$$R_p = \left| \frac{n' \cos I - n \cos I'}{n' \cos I + n \cos I'} \right|^2 .$$

Show that when the reflected intensity is zero the directions of reflection and transmission are at right angles and

$$\tan I_B = n'/n.$$

(This angle I_B is called Brewster's angle; the result is the basis of many polarization devices.)

6 A ray meets a plane mirror at an angle of incidence θ; if the mirror is turned through an angle α about an axis perpendicular to the plane of incidence, show that the reflected ray is turned through 2α.

7 Two plane mirrors are fixed together with their normals intersecting at an angle β. Show that a ray in the plane containing the normals is turned through a fixed angle 2β after reflection at both mirrors, irrespective of its angle of incidence on the first mirror.

8 A mirror has the shape of a conic section of revolution with the axis passing through both foci. Show that a ray through one focus is reflected to pass (possibly virtually) through the other focus.

9 Two positive thin lenses each of focal length f are spaced $2f$ apart. Show that this is an afocal system and calculate the longitudinal magnification. Where must an aperture stop be placed to make the system telecentric?

10 Show that the distance between two real conjugates of a thin lens cannot be less than four times the focal length (the pathological case when both conjugates are at the lens is excluded).

11 An object of size η is projected on to a screen at a fixed distance from the object. Show that there are in general two possible positions for the projection lens and that $\eta^2 = \eta'_1 \eta'_2$ where η'_1 and η'_2 are the sizes of the images at these two positions.

12 [Write a raytracing program for any computer that is available.] A singlet lens has radii of curvature 50 mm and -1000 mm, centre thickness 15 mm and refractive index 1·523. Trace rays from the axial point at -1000 mm from the first surface and find

(a) The transverse ray aberration at the paraxial focus for a ray incident at $y = 21$ mm on the first surface (1·55 mm).

(b) The incidence height on the first surface at which a ray from the same object point has less than 0·005 mm of transverse ray aberration referred to the paraxial focus (3 mm).

13 The lens in example (12) has an aperture stop 20 mm in diameter in contact with the second surface. If the object field at -1000 mm is 100 mm in diameter calculate the Lagrange invariant H. How does the Lagrange invariant change in magnitude if the stop is moved to the right a distance of 50 mm without changing its diameter?

14 Given a paraxial ray traced between an object and an image, both at finite distances, show how the Seidel difference formulae can be used to find the positions of the principal foci. (Eqn (6.16) should be used with an appropriate approach to a limit.)

15 Use the optical sine theorem to calculate the transverse magnification for the finite ray traced at an incidence height of 21 mm in example (12). Compare this with the results of principal ray traces for object heights up to 50 mm, taking the aperture stop at the first surface (the comparison should be made at the paraxial image plane and at the image plane for the finite rays; the differences in magnification are about 7% and 1% respectively).

16 Write down expressions in terms of the variables defined in eqn (7.35) for the fifth order or secondary aberrations. For 5th order spherical aberration calculate the compensating amount of (a) primary spherical aberration to give zero transverse ray aberration at the edge of the pupil and (b) the amount of defocus needed for the same. Plot these in the style of Fig. 7.15.

17 Calculate the Seidel aberration contributions for the lens of example (12) for an object height of 50 mm and for a 40 mm diameter stop in contact with the first surface. $(S_I = 0.458 \text{ mm}, \quad S_{II} = -0.020 \text{ mm}, \quad S_{III} = 0.012 \text{ mm}, \quad S_{IV} = 0.007 \text{ mm}, \quad S_V = -0.0009 \text{ mm}.)$

18 A meniscus lens of refractive index n has a positive first radius of curvature r_1 and centre thickness d. Show that for zero S_I and for infinite object distance the second radius is given by

$$r_2 = \frac{n}{n^2 - 1}(nr_1 - (n - 1)d).$$

A meniscus has zero S_{IV}. Show that for zero S_I with an object at infinity the following condition must hold:

$$r = n(n - 1)d.$$

19 The material of a thin lens has refractive index n and thermal expansion coefficient α. Show that the fractional change of power with temperature T is given by

$$\frac{\delta K}{K} = \left(\frac{1}{n - 1} \frac{dn}{dT} - \alpha \right) \delta T.$$

Appendix D Tracing Gaussian Beams from Lasers

In Section 2.1 it was remarked that the geometrical wavefronts do not exactly coincide in curvature with the physical phasefronts near the focal region and a Gaussian laser beam is a particularly simple example of this effect. The effect is most marked in beams of small convergence angle and since such beams are used extensively in optical systems we give without proof the method of tracing them paraxially through refracting and reflecting surfaces.

A Gaussian beam such as is produced by a helium–neon laser has a position along the axis where the phasefront is plane and the beam has the smallest diameter; this is known as the beam waist and it corresponds to the focus in geometrical optics. At this point the complex amplitude is given by

$$A = A_0 \, e^{-r^2/w_0^2}, \tag{D.1}$$

where r is a radial coordinate, w_0 is the radius of the beam waist and A_0 is the complex amplitude at the centre of the beam. Thus at the beam waist the intensity decreases to $1/e^2$ of its central value at radius w_0.

Figure D.1 shows how the beam propagates beyond the waist; at a distance z along the axis the phasefront has radius of curvature R and the intensity falls to $1/e^2$ of its central value at radius w. However, the beam profile remains Gaussian. The relations between these quantities are given by H. Kogelnik and T. Li, *Applied Optics*, **5**, 1550–1567 (1966), and we reproduce them here.

Starting out from the beam waist we have, at a distance z,

$$w(z) = w_0(z)\left\{1 + \left(\frac{\lambda z}{\pi w_0^2}\right)^2\right\}^{\frac{1}{2}}, \tag{D.2}$$

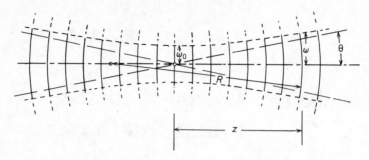

FIG. D.1. Phasefronts in a Gaussian beam. The parameters w_0, z, R and w are shown and it can be seen that at any point along the beam the radius of curvature R of the phasefront is always greater than the distance z to the beam waist

$$R(z) = z\left\{1 + \left(\frac{\pi w_0^2}{\lambda z}\right)^2\right\}, \tag{D.3}$$

where λ is the wavelength in the medium.

Alternatively, to find the beam waist starting from another point along the axis of given R and w,

$$w_0 = w\left\{1 + \left(\frac{\pi w^2}{\lambda R}\right)^2\right\}^{-\frac{1}{2}}, \tag{D.4}$$

$$z = R\left\{1 + \left(\frac{\lambda R}{\pi w^2}\right)^2\right\}^{-1}. \tag{D.5}$$

The beam contour $w(z)$ is a hyperbola of (small) angle θ given by

$$\theta = \lambda/\pi w_0. \tag{D.6}$$

To trace a Gaussian beam through a system we would usually start from the beam waist (this is often at the output mirror of a laser) and find the phasefront curvature at a distance z corresponding to the position of the first optical element, say a thin lens, using eqns (D.2) and (D.3). Next we find the change in this phasefront curvature by means of the ordinary Gaussian optics formulae of Chapter 3. Then we use eqns (D.4) and (D.5) to find the position of the next waist, and so on.

We notice from the equations and from Fig. D.1 that, as mentioned in Section 13.3, the centre of curvature of a phasefront does not lies at the beam waist and the formulae show that the discrepancy is greatest for beams of small convergence angle, i.e. small angle θ of eqn (D.6).

There are, unfortunately, no corresponding simple formulae for dealing with beams of small convergence angle (low Fresnel number) from hard-edged pupils and it would be necessary to carry out a lengthy diffraction calculation to find the difference between the centre of curvature of the phasefront and whatever is taken to correspond to the beam waist.

Name Index

Subject Index